Global Perspectives on Stem Cell Technologies

"Stem cells circulate the globe today in settings where patient desperation confronts regulatory confusion and commercial expectation bumps up against ethical ambiguity. In Bharadwaj's fearless editorial hands, the intersection of these forces, centering on novel Indian therapies, comes alive through voices ranging from academic and reflective to passionate and deeply personal. The book may haunt, destabilize or challenge. It will not bore."
     —Sheila Jasnoff, *Pforzheimer Professor of Science and Technology Studies at Harvard University's John F. Kennedy School of Government*

Aditya Bharadwaj
Editor

Global Perspectives on Stem Cell Technologies

palgrave
macmillan

Editor
Aditya Bharadwaj
Department of Anthropology and Sociology
Graduate Institute of International and Development Studies
Geneva, Switzerland

ISBN 978-3-319-63786-0   ISBN 978-3-319-63787-7   (eBook)
https://doi.org/10.1007/978-3-319-63787-7

Library of Congress Control Number: 2017954290

Cover illustration: © Henry Petrides

Printed on acid-free paper

This Palgrave Macmillan imprint is published by Springer Nature
The registered company is Springer International Publishing AG
The registered company address is: Gewerbestrasse 11, 6330 Cham, Switzerland

Foreword: *Good Science, Better Patients*

In my 2013 book, *Good Science: The Ethical Choreography of Stem Cell Research* (MIT Press),[1] I characterized the current era as one in which two kinds of "good science" go together to drive highly capitalized biomedical innovation: "good science" as in reproducible, reliable, and robust method and knowledge, and "good science" as in freighted with ethical questions such as the moral status of embryos or how to compensate egg donors or find cures. I investigated this intertwining of the moral and technical in bio-innovation economies through the example of US and California pluripotent stem cell research during the 15 years from the first successful derivation of human embryonic stem cell lines until the political and technical stabilization of induced pluripotent stem cell research. I developed a mixed ethnographic/archival/participatory method I called "triage" to collect data with the intent of bringing to light processes whereby some lives come to matter more than others in relation to an emerging technology. I examined US Democrats' and Republicans' competing framings of stem cell research; stem cell research's geographies and geopolitics as seen from California, especially federal and state dynamics in the USA; putative brain drains to the UK and Singapore; and a short-lived rivalry with South Korea over somatic cell nuclear transfer cloning. I also looked at novel public–private funding and governance structures that were being erected for dealing with the ever-present risk of market failure around human cellular technologies at this time, and at how poorly the technical and moral challenges of stem cell research were addressed by the post-war ethics of what I called the "substitutive research subject."

To demonstrate the way that ethical and technical challenges were articulated together throughout research, and not just before beginning the research and after the research is let loose upon the social sphere to have implications or impact, I highlighted California's "procurial" frame for human pluripotent stem cell research: *pro-cures* as the overwhelming moral mandate and technical challenge; *procurement* as the simultaneous ethical roadblock and technical feedstock; and *bio-curation* as the process of moving (de)identified characterized tissue and bio-information between formats and in accounted-for chains of custody. In order to achieve both epistemologically and morally good science, I argued for upstream, interactive, and iterative technical, economic, and ethical innovation, expressed in terms that make sense within the political repertoire of a given jurisdiction. I also argued for the iterative, participatory, and upstream inclusion of distributive justice goals so as to disrupt the systems-entropy whereby discrimination characteristic of a history, time, and place is materialized into technologies and their corresponding ethical instruments. I pushed against the blindness to stratification of the individualism of much of bioethics, and against the species and mind-over-body exceptionalism of humanism. I also called for a radical updating of the epistemologically entrenched and globally circulating and differentiating category of the substitutive research subject. This work has had concrete outcomes, ranging from invitations to teach animal research ethics to engagement with human rights lawyers, and the launch with colleagues of the Science FARE (feminist, anti-racist, equity) initiative to urge technical infrastructures to embed social justice goals.[2]

This important edited collection, *Global Perspectives on Stem Cell Technologies*, takes up good science in ways that resonate with my own development of the term, as well as in quite other ways. In close kinship, this collection focuses on connections forged by cellular technologies through "the twin processes of extraction and insertion of biogenetic substance across multiple terrains ranging from geopolitical borders to areas between biology and machine, governance and ethical dilemmas, everyday suffering, and religious as well as secularized morality," ("Introduction:

Stem Cell Intersections: Perspectives and Experiences") that Aditya Bharadwaj felicitously refers to as "bio-crossings." Nayantara Sheron and Bharadwaj argue in "Bio-crossing Heterotopia: Revisiting Contemporary Stem Cell Research and Therapy in India" that "science and emerging political economy of stem cell technologies around the globe are producing distinct culture-specific responses" (introduction). Sheron and Bharadwaj compellingly reimagine stem cells in India "as heterotopias: manifest entities and discursive sites suffused with real and imagined, and utopic and dystopic alterations made manifest as bio-crossings gain traction between the biogenetic, technoscientific, socioeconomic, and geopolitical landscapes of possibilities."

Sheron and Bharadwaj demonstrate, as I did for the American Medical Association's guidelines on medical tourism, how two sets of 2013 Guidelines in India ended up with the consequence of an "automatic 'good/permissible' science versus a 'bad/rogue' science," between human embryonic stem cells (hESC) and somatic autologous cells. In their chapter, "Staging Scientific Selves and Pluripotent Cells in South Korea and Japan," Marcie Middlebrooks and Hazuki Shimono show how important the genre of scientific biography is to establishing good science, which cannot happen without good scientists. As I did in *Good Science*, they reconstruct the portrayal of South Korea's Hwang Woo-suk before and after the somatic stem cell cloning scandal. They go further and fruitfully compare portrayals of Hwang with portrayals of Japan's Haruko Obokata during and after the stimulus-triggered acquired pluripotency (STAP) scandal. When Hwang and Obokata were national heroes, traditional biographical vignettes were afforded great nationalist significance. Obokata's grandmother's apron in the place of a lab coat, and Hwang's cow with whom he shared bucolic roots as a boy, assured their goodness as moral scientists. Once the scandals broke, however, these same "moral maternal" tropes began to be treated as evidence of the very flaws that led to their downfall.

Where I focused on the problems of extending the ethics and epistemology of substitutive human and animal models to regenerative medicine, Linda Hogle ("Ethical Ambiguities: Emerging Models of Donor–Researcher Relations in the Induced Pluripotent Stem Cells") draws our attention to a different ethical and epistemological limitation emerging in stem cell research: the changing nature of trials and how they advance evidence-based medicine. She draws on her own previous work to show that "for cell-based products, large-scale trials pose challenges, blinding is virtually impossible, and endpoints are difficult to establish." She convincingly argues that the previous gold standard of good science, the double-blind randomized controlled trial, is being superseded by patient activism and computational tools in an emerging assemblage of evidence-based medicine for the pro-cures era.

Sarah Franklin ("Somewhere Over the Rainbow, Cells Do Fly") adds a vital element to the epistemological and geopolitical landscape by zeroing in on regenerative medicine and its associated industries in the UK. As she shows, stem cell technologies are increasingly powerful and "disruptive" because they are part of the "technological platform that enabled the reprogramming of reproductive biology." "Stage 3 clinical trials in a wide range of fields," and "combination products that integrate cells with medical devices, such as patches and scaffolds," lead to "a far-ranging vision of induced plasticity delivered through signalling factors extracted from pluripotent cells and repurposed to trigger in situ cellular reorganisation." Franklin points to these technical breakthroughs as well as to smaller yet higher impactful by-products of changing practice. For example, the trend to freeze embryos at a later stage in fertility medicine has meant that scientists have a diminished source of leftover embryos suitable for hESC derivation. Similarly, the development of stem cell product derived patches that lose many of the cellular properties is making the delivery of cellular products easier. Franklin challenges readers to consider what contribution the social sciences might play. If there are social aspects of regenerative medicine throughout the research, development, and application process, downstream models of social science impact are likely wrong, too. She argues that we need to develop equally sophisticated models to measure and promote our own research impact.

I want to now turn to another fascinating aspect of this edited collection. In the title to this Foreword, I have gestured at it with the phrase, "better patients." This is a play on the moral and epistemological work patients and their physicians and advocates do[3] which makes the science better and thus makes their treatment work better, and on the sense of "being, feeling, or getting better" that we apply to patients who are on the road to recovery or have reduced symptoms and/or improved quality of life. The remaining chapters in this collection concern the kinds of good science captured by this double meaning of "better patients." Dr. Geeta Shroff's chapter, "Establishment and Use of Injectable Human Embryonic Stem Cells for Clinical Application," documents her reasons for choosing a particular population of chronic spinal cord injury (SCI) patients for hESC-derived transplantations in her Indian embryonic stem cell transplant clinic. She focuses on the needs of those "less likely to suffer opportunity costs from study participation" for whom there are few treatment alternatives, and rightly casts this as its own moral imperative. Shroff displays the methodological virtues of her protocol: its simplicity, the regenerative nature of the cell line she uses, and the cell culture; freedom from animal products. The chapter takes time to demonstrate the treatment's scientific legitimacy in highly recognizable terms, but the emphasis is on small clinical improvements that contribute to wellbeing and more productive life, however manifested, rather than on cures or increased survival. Petra Hopf-Seidel, in her chapter "Pre-blastomeric Regeneration: German Patients Encounter hESC in India," likewise explains that "each of my patients improved in one way or other, some visible and measurable, others more invisible with improved stamina, better moods, or more muscle strength. No one had adverse effects, so I can say confidently that I was surrounded by happy patients." Better treatments are ones that lead to better patients however measured.

Finally, Ripudaman Singh's chapter, "Active Parents, Parental Activism: The Adipose Stem Cell In-Vitro Lab Study," and Lola and Shannon Davis' chapter, "Accidental Events: Regenerative Medicine, Quadriplegia and Life's Journey," take us to the heart of the work done by patients and their parent activists to bring treatments in to being in ways that promote good science and lead to better patients. Singh, his wife, and son all worked tirelessly with their fellow families to improve their children's conditions of life in a country where "nobody cares," least of all the Indian Medical Council, which took the dismissive attitude that "stem cells are just placebos." The "parents did most of the research and determined the protocols, such as how many cells we wanted and the number of infusions," and, at the very least, knowledgeably and actively bought "time until something better came along." In their chapter, Shannon and Lola

Davis emphasize the progress made through being a patient of Dr. Shroff when nothing else had worked. Shannon's mother describes the range of incremental improvements in function and quality of life, rather than total cure, offered by the treatments, and sums it up by noting that "her life has become as normal as it can be." Shannon herself underlines the "rigorous medical attention to treatment protocol," as a "deciding factor" in traveling to India for treatment, and also sets the bottom line in terms of efficacy of the treatment in the absence of alternatives, noting that "there is no other place in the world to help me."

Overall, *Global Perspectives on Stem Cell Technologies* was a pleasure to read and learn from. It resonated deeply with my own work, while also taking me much beyond. Good stem cell science and regenerative medicine has much in common around the world, but also crucially differs according to local political and ethical and scientific repertoires, economic circumstances, governance and regulation or the lack thereof, and the institutional structure and funding of science. Narratives of moral and epistemological goodness are produced in and in turn produce scientific and biomedical innovation. Market failures and a completely new understanding of the biological are leading to innovation stretching from clinical trials to patient activism. Cures and care promise, eventually, to be the better for it.

Chancellor's Professor, UC Berkeley;   Charis Thompson
Professor, LSE.
June 2017

Notes

1. Charis Thompson, 2013. *Good Science: The Ethical Choreography of Stem Cell Research*. MIT Press.
2. See *Nature* 538, 371, 2016, Science FARE http://www.nature.com/nature/journal/v538/n7625/full/538317b.html
3. See, for example, these founding works: Steven Epstein, 1998. *Impure Science: AIDS, Activism, and the Politics of Knowledge*. University of California Press, and Rayna Rapp, 1987. Moral Pioneers: Women, Men and Fetuses on a Frontier of Reproductive Technology. *Women Health* 1987;13(1–2):101–116.

Contents

List of Figures

List of Tables

Stem Cell Intersections: Perspectives and Experiences

Aditya Bharadwaj

Introduction

Global Perspectives on Stem Cell Technologies is an exploration of social science, patient, and biomedical perspectives on stem cell technologies. This unique engagement takes as its starting point a humble cell lying on an intersection of ideas as diverse and interlaced as life, knowledge, commerce, governance, and ethics. While natural sciences have focused on the bio-anatomy and unique therapeutic promise of stem cells, social science disciplines such as anthropology and sociology in large part endeavor to reveal the 'cultural contours of interlocked sociotechnical assemblages framing stem cell isolation, generation and application' (Bharadwaj 2012, p. 304). These are shown to range from scientific production, political contestations, and economic calculations to ethical variations, religious objections, and social mobilization around the globe (ibid.). These complex processes and relationships have not only amassed around the scien-

A. Bharadwaj (✉)
Department of Anthropology and Sociology, Graduate Institute
of International and Development Studies, Geneva, Switzerland

© The Author(s) 2018
A. Bharadwaj (ed.), *Global Perspectives on Stem Cell Technologies*,
https://doi.org/10.1007/978-3-319-63787-7_1

tific possibility of purging the cellular form of therapeutic promise but also increased levels of promissory hope and indorsed hype in the cellular form.

This book is an engagement with an emerging but vital area of science spanning geopolitical, socio-economic, and techno-scientific as well as bioethical dimensions. The endeavor is to deepen our understanding of stem cell entities and the concerns, hopes, and aspirations that shape them and make them imaginable as viable therapeutic entities. 'Several key intersections between individual, group, and institutional relationships have become central to locating and debating the production of stem cells' (Bharadwaj 2012, p. 306). Gradually, stem cells are emerging as biogenetic objects bestriding intersections as diverse as ethical/unethical, science/commerce, religious morality/secular governance, somatic/embryonic through to utopian hope and dystopian despair. There is, however, a paradox at the core of stem cell intersectionality: stem cells can be imagined and materially deciphered across a variety of sites. That is, the culturally ascribed identity of stem cells acquires value precisely because stem cells can be imagined as 'both like and not like human beings' (Squier 2004, p. 4). It is on the precise intersection of shifting individual, group, and institutional relationships that stem cells continually renew to mean different things and embody different moral, ethical, economic, and therapeutic values.

The millennial turn saw the rise of the biotechnology of stem cells in nations of the 'South' such as India and beyond. The rapid globalization of stem cell research and clinical application is producing an uneven landscape of opportunity to research, regulate, promote, and debate the cellular form. These moves are also rapidly problematizing long-established oppositions of global North/South, First/Third worlds, developed/developing economies, and Western/Eastern cultures (Bharadwaj 2009; Bharadwaj and Glasner 2009). What is to count as local and global is rapidly dislocating. In large part, this also means that the twentieth-century-development discourse that privileged the unidirectional flow of knowledge from the 'global' North/developed to the 'local' South/developing is disintegrating. As long argued, this geopolitical worldview is now both an untenable orthodoxy and an unsustainable project (ibid.). It is in this world order in which twentieth-century geopolitical stability is rapidly shifting and intersecting in ways previously unimaginable that stem cells have begun to proliferate and mutate to body forth culture-specific responses to certain core and contested arenas.

The book addresses three overarching arenas of concern: (1) regenerating the very notion of regulation and ethics, (2) emerging therapeutic horizons, and (3) patient positions. In large part, these concerns have framed the research focus and lived experience of the authors in this book. These concerns are continually 'co-produced', to use Sheila Jasanoff's apt phrasing, to mean different things in different global contexts. For example, as the accounts in this volume show, while there is emerging evidence of growing social and regulatory concerns around stem cell research and clinical interventions from the United States, the United Kingdom, and Japan, stem cell therapies have become firmly embedded as therapeutic practice in global locales like India. Similarly, in some parts of the world, regulatory and ethical concerns are focused predominantly on the clinical manipulation of the embryonic form and sourcing of reproductive gametes for research (Sperling 2013). In some other global locales, the impact of invasive extraction practices to procure such biogenetic tissues and exploitation of vulnerable populations is being framed as a major area of concern (Waldby and Cooper 2010). The global political economy of such biotechnological developments along with the commercial exploitation of future therapeutic possibilities is also causing alarm and mobilization. While the origins and ethical objections to using embryos for stem cell research can be traced back to the religious domain in specific Euro-American formations (Bharadwaj 2009), the global variability notwithstanding, creation of human embryos for fertility treatments and stem cell research alike have become core bioethical subjects as ethical concerns and, to borrow from Sarah Franklin's insightful analysis, are 'built into' new life forms (2003, 2013). The question of ethics covers a spectrum of issues ranging from scandals involving unethical stem cell research (chapter 'Staging Scientific Selves and Pluripotent Cells in South Korea and Japan') to what Clare Williams and her colleagues have shown to be 'ethical boundary work' (Wainwright et al. 2006) within stem cell laboratories and clinical application of stem cells in specific global locales to the ethics of gamete and embryo sourcing for research through stem cells.

The emerging treatment modalities in a globalized research and therapeutic landscape are similarly mired and caught up in the crude but readily available intersection between good and bad science (Bharadwaj 2015). The conversation this book seeks to instigate significantly involves one notable example of an emerging embryonic stem cell treatment modality in India (see chapters 'Establishment and Use of Injectable Human Embryonic Stem Cells for Clinical Application', 'Pre-blastomeric regeneration: German patients encounter human embryonic stem cells in India', and 'Accidental Events: Regenerative Medicine, Quadriplegia and Life's Journey'). In large part, the intent is to let the voices of those most intimately involved in this breakthrough—from the clinician scientist and author of this unique breakthrough to international interlocutors ranging from physicians to patients embodying the cellular therapy—put forward their perspectives. For too long, these voices have been marginalized in bioscience and social science literature as fringe, guileful, or gullible (cf. Bharadwaj 2013a, b, c, 2015). However, as decade-long anthropological analysis has shown that to the purveyors and surveyors of normative 'good science', clinical breakthroughs in India may seem problematic because they perceive human embryonic stem cells (hESCs) to be digressing from what is often seen as adjudicated and consensible science (Bharadwaj 2013a). Nonetheless, hESC interpolations achieved in the Indian clinic amply illustrate ways in which the slow-paced but high-stakes, capital-saturated, Euro-American forays into stem cell research produce structural conditions that allow the tropic notion of 'bad name science' to solidify on the intersection of states, capital, and science (see Bharadwaj 2015). However, we must remain alert and not lapse into a just as radially available and tempting essentialism that could recast the critique of hESC in India, for example, as mere evidence of a 'West versus the rest' mindset. Rather, it seems the politics of life and science (in that order) paint a more complicated portraiture that takes as their rhetoric of persuasion the notion 'first in the West then elsewhere' (see Chakrabarty 2000, p. 6). Let us also be clear that the emerging global intersection of state, science, and capital is bringing together a collation of strange bedfellows. For instance, the emerging regulatory guidelines in India have more in common with the standardized regulatory norms long fantastically fantasized in the Euro-American landscape as establishing a global gold standard in which biogenetic tissue could become normalized as intellectual property, commercial transaction, standardized therapeutic protocol, and normative bioethical compliance. Put another way, these imagined 'tracks' are fast becoming essential for the smooth shuttling of capital-fueled biotechnological locomotion. The emerging binary between the hESCs and somatic cells in Indian regulatory thinking is a fine reflection of this purportedly globally standardized view on human embryonic source of stem cell as inherently unethical, dangerous (cancerous), and difficult to regulate (chapter 'Biocrossing Heterotopia: Revisiting Contemporary Stem Cell Research and Therapy in India'). It seems the very notion of regulation is in a double bind: how to regulate embryonic stem cell proliferation in petri dishes and across the globe and how to regulate (and not proliferate) ethical, moral, and political issues. Yet, hESCs are proliferating in India and attracting patients from around the globe (see chapters 'Pre-blastomeric regeneration: German patients encounter human embryonic stem cells in India' and 'Accidental Events: Regenerative Medicine, Quadriplegia and Life's Journey'). The so-called regulatory vacuum, as some argue (Sleeboom-Faulkner and Patra 2008), is purportedly allowing this proliferation to go unchecked. The reason this collection includes the Indian hESC breakthrough prominently is because the Indian case is quite possibly the only contemporary example in the world where hESCs are being used clinically with accumulating patient data and testimonies that render problematic the spectral fears of dangerous proliferating potential of embryonic cellular form (Widschwendter et al. 2006).

Against this backdrop, the growing movement of people from around the world in search of stem cell therapies becomes yet another emerging arena of concern. Stem cell tourism, as global therapeutic travel is frequently euphemized, has expanded to include India as a major hub. The so-called stem cell tourists are part of the conversation this book seeks to set in motion, and only their voices can best complicate the problematic nature of the 'tourism' euphemism. It would be erroneous to view this as a mere experimental moment in charting the rise of an innovative biotechnology. Instead, this book's main orientation is a belief that no matter how noble our intentions as social science researchers, we cannot truly give voices to people we 'study' be they scientists, clinicians, or patients.

Instead we can merely create conditions for voices to emerge. Taken together, these developments turn stem cells into a 'spectacle ripe for … analysis' (Hogle 2005).

Stem Cell Theory Machine

Stem cell intersections offer a unique opportunity to revisit Galison (2003) and Helmreich's (2011) notion of the 'theory machine' (also see Bharadwaj 2012), that is, 'an object in the world that stimulates a theoretical formulation' (Helmreich 2011, p. 132). Helmreich explains that for Galison, 'networks of electrocoordinated clocks in turn-of-the-twentieth-century European railway stations aided Einstein's thinking about simultaneity'. Similarly, 'animal husbandry provided a theory machine for Darwin' (ibid.). Retooling Galison, Helmreich focuses on theory as neither fixed above the empirical nor deriving from it in any straightforward sense but rather as crossing the empirical transversely (also see Helmreich 2009, p. 23–25). Thus argued, theory becomes at once an abstraction and an object in the world. In Helmreich's formulation, 'theories constantly cut across and complicate our paths as we navigate forward in the "real world"' (Helmreich 2011, p. 135).

Manifestly a humble stem cell is a theory machine par excellence. As a quintessential 'emergent form of life' (Fischer 2003), a stem cell is at once constricted in the specific context of its cultural medium and dispersed as a 'global biological' entity (Franklin 2005). The theory machine potential of a stem cell is thoroughly realized in its cultural capacity to manifest as the progenitor idea that transforms the notion of 'life' as not only emergent but also simultaneously regenerating. It is the regenerating potential of stem cells, both therapeutically and the social, economic, political regeneration such therapeutic promise sets in motion that further complicates the symbiotic and semiotic emergence of a vital concept: life.

As an abstraction and a real object, a stem cell is rapidly becoming vital to the vitality of the emerging notion of life as regenerative and its evolving institutional and structural framing in the new century. One can argue that the stem cell theory machine crosses sharply athwart the empirical terrain of life. This produces complications. In other words, stem cells are abstractions with real-life consequences. The athwart movement of cells through everyday lived complexities that imbricate science and suffering, as well as regulatory necessities and ethical contingencies, can be seen tropically instantiating a 'biocrossing' (Bharadwaj 2008). As a conceptual trope, the notion of biocrossing alerts us to 'crossings' achieved through the twin processes of extraction and insertion of biogenetic substance across multiple terrains ranging from geopolitical borders to areas between biology and machine, governance and ethical dilemmas, everyday suffering, and religious as well as secularized morality (ibid.). A crucially important way to examine these complexities is to become attentive to ways in which biocrossings traverse the heterotopic spaces in which utopian promise and dystopian angst are reflected and refracted (see Foucault 1986; chapter 'Biocrossing Heterotopia: Revisiting Contemporary Stem Cell Research and Therapy in India', this volume). These reflected sites produce counter-sites within cultures that allow life to assert its vitality within a set of circumstances and material conditions that run counter to individual or shared ideas about life. The theory machine of stem cells is uniquely placed to operate in and as heterotopias: manifest entities and discursive sites suffused with real and imagined, utopic, and dystopic alterations made evident as 'biocrossing gain traction between the biogenetic, technoscientific, socioeconomic, and geopolitical landscapes of possibilities' (ibid.). To be clear, heterotopias are not negative spaces per se but rather multiple concrete and discursive counter-spaces that can be experienced. While Foucault neglected to unpack the notion of heterotopia in any meaningful detail, a close reading of his limited musings on the topic suggests that the notion of heterotopia allows life to unfold and accumulate temporally and spatially even in the face of structural conditions seemingly not conducive to nor sufficient for life. For example, in Foucault's formulation, both prison and museum would typify a heterotopia. While the latter would accumulate time and space indefinitely, the former could become transitory surveyed time and panoptic space. In a similar vein, the temporal and spatial vitality inhered in the cellular form and the vital force of human life itself become equally heterotopic. As counter-spaces, heterotopias contain the potential to operationalize life and enable life to willfully accumulate or dissipate by 'juxtaposing in a single real place several spaces, several sites that are in themselves incompatible' (Foucault 1986, p. 25). These spaces can range from the human corporeal form, stem cells ensconced in a petri dish, hospitals, and laboratories to conference halls, classrooms, and national parliaments promoting or neglecting panoptic ethicality through to international stock markets and pharmaceutical corporate entities. These sites, incompatible in scale, temporality, and power, are importantly reflected and rendered vibrant as they interact and counteract over time and space to produce dynamic shifting social arrangements that ironically sustain and curtail stem cells. Foucault reminds us that 'the heterotopic site is not freely accessible like a public place' (1986, p. 26). The entry into a heterotopia is either compulsorily overseen (e.g., barracks or a prison) or via rites and purifications. This unique heterotopic character isolates as well as renders accessible a counter-site. The purification of stem cells as ethical objects and shards of hermetically isolated and panoptically surveyed biogenetic tissue (imprisoned in a laboratory) further behooves us to inspect the open and closed character of stem cell heterotopia.

The ethical space framing stem cells has a discursive presence. However, the theory machine of stem cells concertizes the discursive and specializes it to hone and 'home in' on competing social orderings that not only harden to become canonical practices and pronouncements but also end up subordinating ethical practices that materialize in response to mundane encounters with life and living. The ordering of good and bad science, however, makes the moral binary factitious. In Thompson's brilliantly insightful account of ethical choreography surrounding stem cell science, she shows that a truly good science with ethics would do more than conceive best scientific and ethical practices as mere instruments for overcoming ethical barriers to research (or for that matter clinical application). Instead Thompson eloquently argues that:

> ... dissent and assent and other interests in relation to fields of science should be solicited, not shut down by scientists and ethicists and administrators; that criticism of science should open up, rather than shutting down avenues of research; that the process and procedures of ethical inquiry should be honored; and that multiple forums for ethical deliberation should be developed, recognized, and made integral to robust science. (2013, p. 64–65)

Perhaps it is time to embrace and advocate the open-ended nature of ethical deliberations, broadly participatory and somewhat democratic, as emerging cellular potential gets realized and theorized around the globe. The alternative and slightly closed and inward-looking bioethical farming merely bureaucratizes ethics to mean something altogether specific. The true answer is perhaps to be (re)searched on the intersection of these competing ethicalities. The theory machine potential of stem cell and its ethical pluripotency is uniquely placed to achieve and propagate this integration.

Regulating Pluripotency

The global stem cell landscape can be imagined as inherently pluripotent. This inherent pluripotency gives rise to much more than vibrant cellular forms—that is, the science and emerging political economy of stem cell technologies around the globe are producing distinct culture-specific responses. It is as if by virtue of differentiating in divergent cross-cultural mediums, stem cell science has become an arena in need of robust standardized regulation. Yet, the notion of regulation remains a slippery concept in much of the social science scholarship and state response to stem cells these accounts focus on as their empirical base. There is an unwitting assumption that greater regulation would somehow rein in the euphemistic pluripotency from assuming dangerous proportions (Salter 2008; Patra and Sleeboom-Faulkner 2009).

Sheila Jasanoff shows that 'biotechnology politics and policy are situated at the intersection of two profoundly destabilizing changes in the way we view the world: one cognitive, the other political' (2005, p. 13). Science has historically maintained its legitimacy by cultivating a careful distance from the politics (Jasanoff 2005, p. 6). She argues that as state-science relations become more openly instrumental, we can reasonably wonder whether science will lose its ability to serve either state or society as a source of impartial critical authority (p. 6). In other words, Jasanoff (1990, 2004, 2005) equips us to ask how inventions, both scientific and social, relate to public and private actors in (predominantly democratic) nations and assist in the production of new phenomena through their support for biotechnology and how they reassure themselves and others about the safety of the resulting changes—or fail to do so (2005, p. 6). Broadly speaking, the notion of 'pluripotent stem cell' encapsulates this troublesome complexity. The issue of unregulated invention and science with its normative inversion—compliant and adjudicated science—circumscribed by state-science consensus in public and private realms produces a shared sense of belonging to an epistemological and regulatory technology. The technoscientific act of honing cells co-produces (Jasanoff 2004) the equally complex task of honing the technoscientific procedure itself. Similarly, the act of reassuring selves and others becomes a manifestly political act of forging a consensual polity of instrumental and ethical action. Moves to standardize and universalize ethical and epistemological procedures are intimately connected to such impulses interested in honing the pluripotent potential of stem cells.

Regulating the social and scientific pluripotency in a globalized research and therapeutic system is a complex task. In the late twentieth and early twenty-first centuries, these moves have birthed the triumvirate of state-science-capital. Increasingly, this troika works to contain, curtail, and cultivate zones of consensible epistemology, shared ethicality, and commercial viability (see Bharadwaj 2013a)—as if anything proliferating outside this consensible vision of a globalized stem cell terrain becomes, like stem cells themselves, peripherally dangerous. The failure to coax cells, science, and society into an orderly development becomes a failure to foresee and prevent a malignant disruption. However, it would be erroneous to assume that some monopolistic state-science machine of global domination is circumscribing stem cells from proliferating 'unregulated' in nation-states and petri dishes. On the contrary, it is becoming increasingly difficult and complex to determine how democratic nations function and respond in the context of the emerging global politics of science and technology around stem cells. For example, Sperling's rich ethnography on the bioethics debate in Germany offers a peek into the established presence of a pronounced sense of 'German' and 'un-German' modes of doing stem cell research (Sperling 2013). The boundaries around German research at best remain ambiguous even as bioethicality posits research inside and outside Germany by German scientists or research on stem cell lines imported rather than indigenously developed as the threshold for precarious border [bio]crossing of the ethical terrain. The Euro-American terrain is internally diverse and distinct. The national cultures of stem cell research and regulation do depart on occasion significantly. However, regulatory protocols and bioethical thinking in the Euro-American formations, differences, and digressions notwithstanding share a distinct philosophical and ideological provenance. While these manifest differently in different nation-states, for example, at the level of the European Union as opposed to individual member states, they do pose problems, as they travel globally. In India alone one finds that while stem cell scientists effortlessly incorporate Western biomedical training and biotechnological developments into their indigenous stem cell tool kits, they do struggle to make sense of normative injunctions around ethics and new regulatory concerns around human embryonic forms. The resounding pushback observed for over a decade can simply be paraphrased to read that the human embryonic form is neither a religious nor a moral nor ethical 'hot potato' in India. Yet, the moves by the Indian state to problematize the destruction of an embryo as an ethical concern, the creation of hESC lines as inherently perilous, and the regulation of such embryonic entities as exceedingly complex reflect the consensus in the Euro-American formations on the subject. More important, the emerging regulatory concern of the Indian state is seeking to transform the stem cell terrain in India by stemming the therapeutic viability of the pluripotent embryonic cell while proactively coaxing the proliferation of autologous cellular research and therapies (see chapter 'biocrossing Heterotopia: Revisiting Contemporary Stem Cell Research and Therapy in India'). Manifestly, it is no surprise that the emerging stem cell nations like India are seeking to create global reach and access by co-opting and building into the stem cell entities ethical, moral, and regulatory thresholds of their probable lay and professional consumers and future markets (see Bharadwaj 2009). The triumvirate of state-science-capital necessitates that political regulation, scientific consensus, and economic calculation seamlessly align if nascent entities like stem cells are to become viable as ethical, therapeutic, and commercial objects. To read these emerging socio-political complexities as mere standardized regulatory and bioethical practices or in some unique sense hallmark good science would be hugely one dimensional.

Policy and regulatory thinking that assumes simplistic divisions such as good/bad and ethical/unethical often miss the nuanced complexities routinely imploding such binaries. If we subject prefixes such as 'good' and 'bad', usually appended to an idea of science, to critical scrutiny, we soon discover that these prefixes curiously circulate and mutate as they converse with their immediate and distant 'environments' and in so doing attach and detach from the very idea of 'science'. Take, for example, the controversy surrounding Proposition 71 of 2004 (or the California Stem Cell Research and Cures Act), a law enacted by California voters to support stem cell research, most notably embryonic stem cell research, in the state. The California Institute for Regenerative Medicine (CIRM) became the state agency brought into existence by the passage of Proposition 71. Funded by state bond funds and backed by taxpayers to the tune of three billion over ten years, the CIRM became a unique holding space for hype/hope, promise/despair, risk/reward, and intractable diseases/promissory cures (Bharadwaj 2015, p. 4). However, the promissory value of the CIRM was somewhat tarnished when local media began highlighting its 'insular' and 'insider-like' way of doing business (*Los Angeles Times* 2014). The main bone of contention was the CIRM's former president's unethical practices and the subsequent CIRM-sponsored cover-up. From its very inception, the CIRM was to be the crucible of good science, and its remit was to find cures for humankind's worst afflictions. This 'procurial' remit, to use Charis Thompson's felicitous framing, was the defining feature of the CIRM's rapid and unprecedented rise. However, the 'procure' rhetoric of 'good science' that enabled the CIRM to come into existence in the first place paradoxically bore fruit in distant India. The fact of stem cell therapies in India can achieve and deliver results that elude good science elsewhere remains an enduring irony. This is because the critique often encountered in the Indian stem cell terrain has in large part focused on imagined violations of an epistemic kind: no animal models or clinical trials and/or no standardized ethical choreography prefiguring good scientific performativity. In this respect, following Shroff's work (chapter 'Establishment and Use of Injectable Human Embryonic Stem Cells for Clinical Application') is illuminating in one crucial respect: it lays bare the pursuit of 'local good' circumscribed by contingent ethics produced in relation to sensibilities populating the everyday engagement with life (see Das 2015). For instance, in all my interactions with Geeta Shroff, I have found her to see placebo-controlled trials as unethical since stem cells at her clinic are used to treat only terminal and incurable conditions:

> We never opted for a clinical trail because we are against giving placebos. The patient is the control because there is chronicity, and it is not fair to treat a patient with placebos especially if a motor-neuron-disease patient is coming to you who is going down every day. The institutional ethics committee took this decision a very long time ago that there will be no placebo, as it is against our ethics; we can't stand back and watch a motor-neuron-disease patient rapidly worsen and die. It is against our ethics. (Bharadwaj 2015, p. 13)

How do we then accommodate this call for localized ethical contingency in the grand narrative of bioethics? In the register of everyday ethics that Veena Das (2015) has brilliantly illuminated through her work, the contingency and frailty of the human condition and its unpredictable social trajectory render untenable a scientific and bioethical commitment to standardized epistemic choreography. However, procedures and processes are changing. As Hogle shows within the purview of the Twenty-First-Century Cures Act in the United States, the law is instructing the FDA in no uncertain terms to use observational data in the evaluation of drugs, biologics, and devices. This data, Hogle explains, could come, in addition to other sources, from case histories and patient narratives about their own experience (chapter 'Ethical Ambiguities: Emerging Models of Donor–Researcher Relations in the Induced Pluripotent Stem Cells'). While these moves stop far short of a watershed moment in eliciting evidence, newer and older notions of appropriate evidence are likely to become more hybrid (ibid.). Nevertheless, these developments can only give hope. For now, it seems, the mode of building and doing 'good science' as envisioned by Thompson seems a step closer to realization.

On the question of regulation, certain expedient logics appear to underscore the rise of science policy and governance around the globe today. This expediency, I think, is an unwitting corollary (and on rare occasions a willful manifestation) of processes that both operate and are operationalized as the global circulation of intellectual and monetary capital gain traction. We need to pay particular attention to such an emergence within the policy landscape, national and regional differences notwithstanding. We should also remain somewhat ambivalent in the face of two popular and explicit suggestions embedded in the existing social science literature on stem cells that see robust governance of stem cells predicated on common acceptable principles and mechanisms as facilitating good scientific practice and international collaborations and the standardization and globalization of ethical concerns. In my view one of these aims, international collaborations encouraging good scientific practice, is often unattainable given the woeful lack of a level global playing field; the other, the standardization of ethical concerns, is undesirable. This is because in order to understand science policy and regulation, we also need to understand how power structures set definite limits to individual and collective negotiating capacities. The resulting negotiating choreography produces seemingly new norms, but these reassert the hegemonic view that either seeks to co-opt the emerging new in its own image or reject it altogether, a sense of 'our way or the highway'.

The foregoing policy, scientific, ethical, and regulatory concerns often eclipse one important stakeholder in the global stem cell landscape: patients suffering from chronic and degenerative medical conditions. Ironically, the manifesto of 'good science' that Thompson troubles and expands to include a diverse pool of concerns and ethicalities takes as its point of departure a strong 'pro-cure' stance as the main justification for intensified research, enhanced funding, and procuring access to biogenetic tissue. The affect saturated call for this intensification takes human suffering and progressive and degenerative afflictions as the only humane justification for developing and delivering therapy-grade stem cell technologies. The suffering patient thus co-opted in the triumvirate circuit of state-science-capital paradoxically serves to obfuscate the troika at the cost of her own obfuscation. The suffering patient and her suffering is deferred, disappeared, and dispersed into a promissory therapeutic future. The certainty of her suffering and eventual end in the present assumes a totemic quality: a sacrifice that guarantees promised future returns on the investment elicited in her name from state, science, and capital.

I have had the rare privilege of documenting and following biographies of stem cell treatment seekers for nearly a decade. I am delighted that rather than represent them, some of these inspirational pioneers will represent themselves and their experiences in the pages of this book. As noted previously, it is my firm belief that no matter how noble our intention as researchers we cannot truly give voices to people. Instead we can merely create conditions for voices to be heard.

Through the course of my research, I have encountered numerous patients reporting reversals in their rapidly worsening conditions post-stem cell interpolations and voicing deep frustrations on being seen as either psychosomatic or responding to mere placebos (Bharadwaj 2013b). For example, many patients had to contend with well-meaning but unsupportive biomedical opinions advising against stem cell treatments in India. Patients were continually asked to wait for therapeutic alternatives to emerge within their home countries in Europe or the United States. The well-meaning tropic construct of desperate gullible dupe in need of protection from a guileful maverick often silenced the enduring frustration patients articulated. To these intrepid treatment seekers, the ethical stance of principled good science seemed callous and inhuman. As one treatment seeker told me, 'They [purveyors of bioethically settled stem cell science] appear to be saying we rather you die than try'. In a similar vein, a young man told Thompson (2013) he would travel abroad for stem cell treatments if he could. He couldn't understand why there were concerted efforts to demonize countries offering treatments even if those interventions were largely experimental. To the young man, the demonized experimental nature of stem cell treatment modality abroad was more desirable than dying waiting for the FDA in the United States (Thompson 2013, p. 16).

It appears the figure of an independent, autonomous, free, rational, and calculating subject—routinely resurrected in ethically adjudicated consent procedures—is rendered problematic, as a decision to seek stem cell treatments around the globe cannot be captured under the sign of a clinical trial or some form of normative treatment seeking. It appears outside the state-science-capital circuit; autonomy, consent, and choice add up to mean something rather specific—gullibility and desperation. Alternatives to what I am calling the triumvirate-sponsored biomedical science are rendered untenable. And yet therapeutic migrations from over 50 countries to India have continued to seek out stem cell treatments for over a decade (chapters 'Establishment and Use of Injectable Human Embryonic Stem Cells for Clinical Application', 'Pre-blastomeric regeneration: German patients encounter human embryonic stem cells in India', and 'Accidental Events: Regenerative Medicine, Quadriplegia and Life's Journey').

In highlighting the complex pieces making up the pattern of global stem cell initiatives, this book is seeking to initiate and invite conversation. The chapters that follow might offer a template for future engagement and forays into the cellular terrain populated by multidisciplinary stakeholders.

The Book

This book aims to instigate conversation. In so doing we need to remain alert and open to asking what kinds of science, politics, and ethicality are at stake as stem cell science and therapies throw roots around the globe. This will entail crossing disciplinary, ethical, geopolitical, and cultural borders. The chapters that follow offer remarkable insights into groundbreaking research from across disciplines. These perspectives reinforce a call for methodological immersion that is longitudinal, sustained, and multi-sited in order to reveal everyday complexities at the heart of these emerging stem cell challenges around the globe.

The chapters that follow offer illustrations into the emerging life of stem cell technologies in an interconnected world. These examples are unique, and given the prevailing contentious bioethical framing of stem cell entities, some of these illustrations may even be perceived as controversial. One of the primary aims of this collection is to jolt us out of our epistemic comfort zones and facilitate a dialogue on a disciplinary and experiential intersection. As noted previously, the book is held together by three distinct and yet connected thematic sets.

The first major thematic group is concerned with the notion of regenerating regulation and ethics. Franklin (chapter 'Somewhere Over the Rainbow, Cells Do Fly'), Hogle (chapter 'Ethical Ambiguities: Emerging

Models of Donor–Researcher Relations in the Induced Pluripotent Stem Cells'), and Middlebrooks and Shimono (chapter 'Staging Scientific Selves and Pluripotent Cells in South Korea and Japan') illustrate the regulatory and ethical precarities as well as glimpses of emerging new stability in vastly different contexts in the United Kingdom, the United States, South Korea, and Japan.

Franklin argues that cell therapy and regenerative medicine are tied to translational ambitions seeking to deliver improved healthcare. These moves often manifest as 'pipeline models' of delivery and congregate around the discourse of 'impact'. Franklin shows that the pipeline idiom is 'inadequate to encompass the iterative, loping, and often circuitous realities of "translating" knowledge into products and applications'. Drawing on longitudinal ethnographic immersion and proactive conversations with cell-therapy advocates and stem cell researchers, she shows how, when discussions of impact are examined alongside humanities scholars, many common themes begin to emerge. Franklin calls for a move away from linear models of progress to incorporate 'churn', 'circularity', and 'conversations' as the 3Cs in the co-produced future of science and social science. In so arguing, she maps out the various 'intersections' between social and basic science. Franklin expertly troubles the irony underscoring 'promotional' and 'aspirational' idioms impeding 'the very flows they are allegedly designed to accelerate'. She argues that good solutions require a much more circular process. In the final analysis, she calls for better models than 'pipelines' and 'impact' to help appreciate the complexity of technological change. Following Franklin, we can argue that the current-event horizon of stem cell science is ironically birthing variegated rainbows. And perhaps if we fly high enough over the rainbow, a globalized consensus on how to culture, restrict, and circulate stem cell biogenetic entities might become realizable.

Hogle delves into the world of stem cell and regenerative-medicine governance. She examines the contemporary debates over regenerative-medicine implementation and governance in the context of emerging thinking on producing evidence in contemporary biosciences and medicine. She persuasively argues that stem cell and regenerative-medicine governance has largely been circumscribed by technological zones and limited to: what is or is not allowed by regulatory authorities in specific locales, what is or is not an ethical therapeutic application, and the variances across societies. She shows how this approach largely ignores intersections with economic, political, and other kinds of technological zones. Hogle makes a ground-breaking intervention by problematizing the category of evidence itself. She shows how stem cells upset stable categories set forth by evidence-based medicine and policy because of their 'complexity and recalcitrance to existing ways of measuring evidence'.

Hogle offers a fascinating insight into the current state of flux where the following are ongoing: a shift toward patient-generated data and patient entitlements to choose experimental treatments; a push to speed up product approvals circumscribed by differing attitudes toward risk and patients' roles in decision-making; an uptake of new techniques such as Big Data analytics and predictive computation that aid economic calculations for systems as a whole well beyond the production of data for specific innovations; and actions built on platforms serving broader political and economic purposes. In this climate of change she rightly impels us to ask what work we are expecting evidence to do in the ethically ambiguous stem cell terrain.

Middlebrooks and Hazuki explore how prominent Japanese and South Korean scientists Obokata Haruko's and Hwang Woo-suk's public personas and self-presentations produced the credibility of their stem cell research narratives. The chapter offers a gripping account of ways in which extensive media coverage of both scientists' stem cell successes and subsequent stem cell research scandals dovetailed their public personas to the 'ontological possibility of their promised stem cells in fluid yet persistently gendered ways'. Middlebrooks and Hazuki argue that the Stimulus-Triggered Acquisition of Pluripotency stem cell research scandal in Japan and the human embryonic somatic cell nucleus transfer or cloned stem cell research scandal in South Korea link the perceived integrity of mass-mediated scientific personas with the 'integrate-ability' of their stem cell research results. The chapter lays bare the vulnerability of ethical and regulatory oversight in the face of stage-managed 'scientific selves' via personalized public performances in sustaining public support for stem cell science.

The second thematic segment takes the reader into the biomedical terrain of human embryonic stem cell innovation in India. Despite much promissory hope and hype invested in therapeutic viability in the Euro-American formations, the Indian example complicates our understanding of stem cell therapies in a globalized research system. Shroff, through her extensive work treating spinal cord injury with hESCs, argues how despite their great potential in curing chronic conditions such as spinal cord injury (SCI), hESCs have not been used extensively in humans. She shows that current research on treatment options for traumatic SCI aims at regaining the lost functions of the spinal cord by promoting re-myelination (material surrounding nerves) with oligodendrocytes (concerned with the production of myelin [an insulating sheath around many nerve fibers] in the central nervous system) and formation of neurons. The case studies detailed in this chapter are the first of their kind to demonstrate the adequate efficacy of hESCs in SCI patients with a good tolerability profile. Shroff draws on accumulated data to show how patients gained voluntary movement of the areas below the levels of injury as well as improvements in bladder and bowel sensation and control, gait, and handgrip. The chapter offers potentially landmark insights into the therapeutic potential of largely misunderstood hESC transplantation in SCI patients.

After seeing a successful hESC case at a conference in Germany, Hopf-Seidel accompanied 12 patients from 20 to 73 years of age with chronic conditions such as Lyme, amyotrophic laterals sclerosis, arthritis, and macular degeneration to India for treatment. Faced with intractable and debilitating conditions in her patients, she recommended pre-blastomeric embryonic stem cell therapy in India. The chapter details the outcome of three intensive trips to the clinic between 2012 and 2014 with patients who could not experience any improvement through previous conventional medical treatments. The chapter traces the journey and illustrates the outcomes based on photographic and biomedical evidence gathered on these trips and subsequent follow-ups in Germany.

The third and final segment takes us into the world of patient positions on stem cells. Singh as well as Davis and Davis show in their respective chapters how these positions offer literal examples of patience and resilience, while Appleton and Bharadwaj draw on patient and practitioner experiences in the larger context of engineered shifts in the Indian policy landscape.

The notion of 'active parent' blurs the lines between parental and professional activism. Singh explores this complex intersection to show how active parents and parental activism intersect to produce a unique biography of an emerging stem cell intervention. The chapter documents the personal journey of Singh as a working professional who took on the seemingly impossible task of finding a cure for his four-year-old son, who in 2005 was diagnosed with Duchenne Muscular Dystrophy, a muscle-wasting condition. The chapter traces the deeply personal account of accepting, resisting, and rejecting the diagnosis and the intractable finality it presented. This account emerges from an autobiographical space and narrates the birth of an 'active parent' who with 10 other 'active parents' (connected to more than 200 parents) took on the challenge of finding an adipose stem cell-based cure. The chapter charts the failures and successes on the path to directing and driving the study and how parents coped with the demands of laying down the complete study protocols through to ensuring the safety and efficacy of the study to secure some semblance of therapeutic value for their children.

When Shannon Davis became quadriplegic after a devastating and life-altering car accident, she sought treatment in India from Dr. Shroff. In the first three months of treatment, Shannon showed improvement in all muscle groups and was able to stand upright with leg and abdominal calipers for longer and longer periods. In this chapter, the Davises argue that while the potential of stem cells to transform medicine will be a reality one day, for families in need of help today (or yesterday), the urgency to make decisions plays a critical role. The account shows how parents of desperately ill or injured children, especially those for whom no established treatment exists, search for and are often willing to engage in treatments in far corners of the world with potential positive outcomes. In the final analysis, they share the process of their travel to India and the experience of receiving positive results via human embryonic stem cell treatment.

Appleton and Bharadwaj show that the fraught and contested terrain of stem cell research and therapies is an undulating landscape of utopias and dystopias. While dystopic scenarios of stem cell research and therapy in unregulated and unregimented nation-states include fear of mass epidemics of cancerous growths in uninsured, ill-informed, and gullible patients, the utopic scenario imagines personalized medicine without multi-national pharmaceutical profit motivations or leading hospitals and physicians acting as gatekeepers for accessible care. Extending the tropic notion of 'biocrossing' (Bharadwaj 2008), the chapter articulates the faint traces of utopic and dystopic logics underscoring these 'crossings' and the evolving biography of a contested terrain this (re)scripts. Appleton and Bharadwaj engage with ethnographic immersion into the lives of physicians, researchers, policymakers, and patients to conceptualize evolving scenarios that remain divergent and yet the source of emergent but shifting utopias and dystopias that often are experienced as a heterotopia.

*   *   *

This book produces a unique account of the emerging research/therapy interface in order to explicate the high-risk and high-gain production of stem cell biotechnologies around the globe. The collection situates these developments in the context of larger global developments, most notably, the United States, Europe, and Asia to excavate the multi-national and multi-sited nature of contentious innovation culturing the stem cell technology landscape. Our hope is to provide an insightful account detailing arenas of stem cell research; local and global trajectories of therapeutic application and scientific collaborations; lines of public- and private-sector intersections; zones of ethical contestation; implications for private- and public-sector investments in science and biotechnology; and the tenuous nature of governance and its implications for both Euro-American science and burgeoning regenerative biotechnology sectors in India. In other words, this book is small but has big aspirations. It's a dialogue across cultures: social sciences and biosciences, Indian science and Euro-American science, clinical scientists providing stem cell care, and patients embodying these scientific breakthroughs. The common denominator is the word 'science': it brings us together, binds us together. While science is curiosity and the pursuit of knowledge and ideas, our points of departure and cultures of practice are deeply informed by how and where we are located: institutionally, culturally, as well as geographically. Much like stem cells and their regenerative capacity, our work practices and thought processes also gestate in a distinct cultural medium. Our sincere hope is that this book will be the starting point of a unique mixing of cultures seemingly removed from each other. It seeks to inaugurate a conversation across disciplinary and national boundaries and share outcomes of research-led understanding and interdisciplinary collaborations. While we remain embedded in our respective cultures of knowing, problem-solving and playing to our inimitable strengths and unique approaches to understanding the cellular form would, I strongly feel, succeed in enabling a shared understanding of what collaborative effort can achieve. It is in this spirit of collaboration and common interest in the cellular form that we ought to attempt moving forward.

References

Bharadwaj, Aditya. 2008. Biosociality and Biocrossings: Encounters with Assisted Conception and Embryonic Stem Cell in India. In *Biosocialities, Genetics, and the Social Sciences: Making Biologies and Identities*, ed. Sahra Gibbon and Carlos Novas, 98–116. London; New York: Routledge.

———. 2009. Assisted Life: The Neoliberal Moral Economy of Embryonic Stem Cells in India. In *Assisting Reproduction, Testing Genes: Global Encounters with New Biotechnologies*, ed. D. Birenbaum-Carmeli and Marcia C. Inhorn, 239. New York: Berghahn Books.

———. 2012. Enculturating Cells: Anthropology, Substance, and Science of Stem Cells. *Annual Review of Anthropology* 41: 303–317.

———. 2013a. Ethics of Consensibility, Subaltern Ethicality: The Clinical Application of Embryonic Stem Cells in India. *BioSocieties* 8 (1): 25–40.

———. 2013b. Subaltern Biology? Local Biologies, Indian Odysseys, and the Pursuit of Human Embryonic Stem Cell Therapies. *Medical Anthropology* 32 (4): 359–373.

———. 2013c. Experimental Subjectification: The Pursuit of Human Embryonic Stem Cells in India. *Ethnos* 79 (1): 84–107.

———. 2015. *Badnam* Science? The Spectre of the 'Bad' Name and the Politics of Stem Cell Science in India. *South Asia Multidisciplinary Academic Journal* 12: 1–18. Accessed April 21, 2017. doi:10.4000/samaj.3999

Bharadwaj, Aditya, and Peter Glasner. 2009. *Local Cells, Global Science: The Proliferation of Stem Cell Technologies in India*. London: Routledge.

Chakrabarty, Dipesh. 2000. *Provincializing Europe: Postcolonial Thought and Historical Difference*. Princeton: Princeton University Press.

Das, Veena. 2015. *Afflictions: Health, Disease, Poverty*. New York: Fordham University Press.

Fischer, M.M.J. 2003. *Emergent Forms of Life and the Anthropological Voice*. Durham, NC: Duke Univ. Press.

Foucault, Michel. 1986. *Of Other Spaces*. Translated by Jay Miskowiec. *Diacritics* 16 (1): 22–27.

Franklin, Sarah. 2003. Ethical Biocapital: New Strategies of Cell Culture. In *Remaking Life and Death: Toward an Anthropology of the Biosciences*, ed. Sarah Franklin and Margaret Lock, 97–128. Santa Fe, NM: University of New Mexico Press.

———. 2005. Stem Cells R Us: Emergent Life Forms and the Global Biological. In *Global Assemblages: Technology, Politics, and Ethics as Anthropological Problems*, ed. A. Ong and S.J. Collier, 59–78. New York; London: Blackwell.

———. 2013. *Biological Relatives: IVF, Stem Cells, and the Future of Kinship*. Durham, NC: Duke University Press. http://www.oapen.org/search?identifier= 469257

Galison, Peter. 2003. *Einstein's Clocks, Poincaré's Maps: Empires of Time*. New York: Norton.

Helmreich, S. 2011. Nature/Culture/Seawater. *American Anthropologist* 113: 132–144. https://doi.org/10.1111/j.1548- 1433.2010.01311.x

Hetherington, K. 1997. *The Badlands of Modernity: Heterotopia and Social Ordering*. London: Routledge.

Hogle, Linda. 2005. Stem Cell Policy as Spectacle Ripe for Anthropological Analysis. *Anthropology News* 46: 24–25.

Jasanoff, Sheila. 1990. *The Fifth Branch: Science Advisers as Policymakers*. Cambridge, MA: Harvard University Press.

———, ed. 2004. *States of Knowledge: The Co-production of Science and Social Order*. New York: Routledge.

———. 2005. *Designs on Nature: Science and Democracy in Europe and the United States*. Princeton: Princeton University Press.

Patra, P.K., and M. Sleeboom-Faulkner. 2009. Bionetworking: Experimental Stem Cell Therapy and Patient Recruitment in India. *Anthropology & Medicine* 16: 147–163.

Salter, B. 2008. Governing Stem Cell Science in China and India: Emerging Economies and the Global Politics of Innovation. *New Genetics and Society* 27: 145–159.

Sleeboom-Faulkner, M., and P.K. Patra. 2008. The Bioethical Vacuum: National Policies on Human Embryonic Stem Cell Research in India and China. *Journal of International Biotechnology Law* 5: 221–234.

Sperling, Stefan. 2013. *Reasons of Conscience: The Bioethics Debate in Germany*. Chicago: University of Chicago Press.

Squier, Susan Merrill. 2004. *Liminal Lives: Imagining the Human at the Frontiers of Biomedicine*. Durham, NC: Duke University Press.

Thompson, Charis. 2013. *Good Science: The Ethical Choreography of Stem Cell Research*. Cambridge, MA: The MIT Press.

Wainwright, S.P., C. Williams, M. Michael, B. Farsides, and A. Cribb. 2006. Ethical Boundary-Work in the Embryonic Stem Cell Research. *Sociology of Health Illness* 28 (6): 732–748.

Waldby, C., and M. Cooper. 2010. From Reproductive Work to Regenerative Labour: The Female Body and the Stem Cell Industries. *Feminist Theory* 11: 3–22.

Widschwendter, Martin, et al. 2006. Epigenetic Stem Cell Signature in Cancer. *Nature Genetics* 39: 157–158.

Aditya Bharadwaj is a professor at the Graduate Institute of International and Development Studies, Geneva. He moved to the Graduate Institute, Geneva, in January 2013, after completing seven years as a lecturer and later as a senior lecturer at the School of Social and Political Studies at the University of Edinburgh. He received his doctoral degree from the University of Bristol and spent more than three years as a postdoctoral research fellow at Cardiff University before moving to the University of Edinburgh in 2005. His principal research interest is in the area of assisted reproductive, genetic, and stem cell biotechnologies and their rapid spread in diverse global locales. In 2013, he was awarded a European Research Council Consolidator Grant to examine the burgeoning rise of stem cell biotechnologies in India. Bharadwaj's work has been published in peer-reviewed journals such as *Medical Anthropology*; *Ethnos*; *BioSocieties*; *Social Science & Medicine*; *Anthropology & Medicine*; *Health, Risk & Society*; and *Culture, Medicine and Psychiatry* and he has contributed several chapters to edited collections. He has co-authored *Risky Relations, Family Kinship and the New Genetics* (2006) and is the lead author of *Local Cells, Global Science: The Proliferation of Stem Cell Technologies in India* (2009). His sole-authored research monograph is titled *Conceptions: Infertility and Procreative Technologies in India* (2016).

Part I

Regenerating Ethics

Somewhere Over the Rainbow, Cells Do Fly

Sarah Franklin

Three of the most striking features of technological change in the twenty-first century are its speed, variety, and scale: the means by which people communicate, shop, travel, work, read, play, associate, and learn have undergone continuous 'disruptive' changes over the past 50 years, and in this half century the rise of television, space exploration, and the airline industry have been succeeded by genetic engineering, computing, and the internet.

Two key intersections can help us explain the rapid pace of technological change affecting all aspects of contemporary society—including health technologies and the process generally known as 'translation' whereby a technology becomes widely used, profitable, and normalised. The first key intersection is the increasing interconnectivity between technological domains that enables radical changes to traditional activities such as agriculture—in which today, routinely, robotics, computing, gene editing, and satellite technology are combined to enable more efficient planting, cultivation, and harvesting of new types of crops. These

S. Franklin (✉)
Department of Sociology, Cambridge University, Cambridge, UK

A. Bharadwaj (ed.), *Global Perspectives on Stem Cell Technologies*,
https://doi.org/10.1007/978-3-319-63787-7_2

27

are the same recombinant intersections that enable unprecedented scales of capacity to be achieved, such as how many planes can be in the air at the same time and how many passengers can buy seats on them using electronic-ticketing apps on their mobile phones. The second key intersection is between technology and subjectivity—or consciousness. For we cannot explain technological change in terms of capacity alone: it must also be explained in terms of social identities, orientations, and associations. There must be a shared perception of a need and a whole worldview that supports this perception in order for machines to be invented that will use complex evolving algorithms to learn to hear your voice and speak back to you or even to drive your car.

One of Karl Marx's most famous quotations can be summarised as the hand-mill gives you feudal society, the steam mill the industrial capitalist (1971, p. 9).[1] What he meant, however, was not simply that a different technology produced a different kind of society. His interest lay in the specific nature of what he understood to be a complex evolutionary process, not unlike the changes to physiology affecting speciation that so interested his contemporary Charles Darwin. Why, Marx wanted to know, was a specific species of technology—the windmill—in operation for millennia before the sudden change to steam power? Why were careful technological adjustments made to windmills by hundreds of generations of mill-dependent societies for centuries prior to the frantic period of technological replacement that suddenly erupted in the late seventeenth century? Marx argues we cannot explain the revolutionary triumph of steam-driven mills over water- and wind-milling as a result of technological capacity, or power, alone because the orientation of innovation has changed so dramatically, which Marx argues must reflect a change in social structure—and above all a change in consciousness, or worldview.

The Bioindustrial Revolution

The question of rapid, or revolutionary, technological development has become one of the most pressing sociological questions before us not only because it is so difficult to explain—but equally because it is so tempting *not* to explain technological change at all. This is because one of the signs a technology has become revolutionary is that it has become obvious, normal, and ordinary: the mobile phone seems to have become successful because so many people want to use it and because it delivers the step-changing functionality consumers are eager to buy. As with the mobile phone, so too with the internet, email, and computing: the utility of each seemingly explains its ubiquity and vice versa. Neither Marx nor many contemporary sociologists would deny the powerful role of basic scientific discovery in the process of technological change. However, there remains a compelling and even self-evident case that discovery alone is not a sufficient explanation for the dramatically increased speed and scale of technological change experienced since the end of the eighteenth century.

According to the classic Marxist argument that social relations determine technological change, and not the other way around, the most important relation is between the capitalist owner of the means of production and the wage labourers who have been disenfranchised not only of their individual labour power and skills, but also of their collective associations with the shared activities of production and their alienation from the means of production. Marx argues that much of the evolution of machinery during the industrial revolution was driven by the motivations of the capitalist-owner class to extract more value from the proletariat—a motivation that increasingly had become a norm and a requirement under the emergent social form of industrial capitalism. The crucial historical turning point of this process and its most revolutionary moment, he argues, is the point at which machines begin to make other machines, which will in turn replace workers with a more easily controllable, more efficient, tireless, and inanimate means of production.

From this point of view, technological change must be understood in terms not only of the production of desired goods, such as commodities, but also in terms of the broad social and historical forces that shape perceptions of value. The extraction of surplus value for profit achieved through increasing managerial control over the labour process is a recent idea: the value of work was not traditionally defined in such a manner. Mass production did not become socially valued because steam engines made large-scale industrial manufacturing possible. To the contrary, the highly specialised division of labour required by increasingly mechanised manufacture was consistently resisted by the urban proletariat, who saw few of its rewards. At the same time, the process of industrialisation has undoubtedly brought enormous benefits and marked the beginning of a new era of rapid worldwide economic growth—so rapid that in only two centuries the world's population has increased tenfold and a new geologic era has been proposed to mark the impact on the earth itself of industrial manufacturing.

Meanwhile, yet another dimension of technological change is upon us, and in our own century, the rise of the biological sciences has raised a very different set of questions about the relationship of new technologies to social consensus and to industrial capitalism. No one has yet written the equivalent of Marx's *Capital* (1867) for 'the age of biology', although much ink has been spilt on the step-changing consequences for the human species of molecular genomics. In fact it may turn out that the new technologies derived from developmental and reproductive biology, rather than the human genome project, provided the equivalent of the historical turning point in the mid-nineteenth century, when machines began to make other machines, for this is when the potency of cells began to be harnessed to make other cells.

Fittingly, it was exactly a century after the posthumous publication of Marx's magisterial three-volume *Capital: A critique of political economy* that Robert Edwards, the Cambridge biologist, phoned Patrick Steptoe, the consultant obstetrician based just outside Manchester, in Oldham, to discuss the possibility of collaborating on a new means of technologically assisted human reproduction, namely, in vitro fertilisation (IVF). Just over a decade later Louise Brown was born, and a second industrial revolution began to unfold in northwestern England—this time based on a technological platform that enabled the reprogramming of reproductive biology. If IVF is the equivalent of the steam engine in what Ian Wilmut, Keith Campbell, and Colin Tudge (2001) have named 'the age of biological control' that is not only because it is so successful technologically (by which measure the steam engine is in a different league altogether) but because it is so popular. Crucially, IVF is successful and revolutionary, because it has been accompanied by a change in consciousness about not only human reproduction, but biology in general. In sum, biology—including human reproductive biology—has come to be seen as a vast tool kit, a technological horizon, a new frontier of scientific and economic growth (Landecker 2007; Franklin 2013). The pursuit of new means to repair, reprogramme, and redesign biological entities—from genes and cells to bacteria and plants, as well as livestock and people—closely resembles the process described by Marx a century and a half earlier of closely interlinked transformations in technology and consciousness affecting the means of production. Today it is the means of reproduction that are coming to be perceived as the promissory source of vast health and wealth benefits.

Today, as in the early nineteenth century, a great acceleration of technological change is occurring in the realm of what we might call biological equipment—including our own as well as the myriad new species of apparatus and instruments manufactured for the biotechnology sector, such as gene sequencers, incubators, and time-lapse embryo monitors. Today, as in the past, the industrialisation of the bio tool follows a well-beaten path from the bespoke, site-specific crafting of basic scientific tools such handmade primers and pipettes to their industrial manufacture and marketing. We are witnessing the gradual expansion, standardisation, and mass production of whole new families of tools such as synthetic antibodies, receptors, blockers, and other cell- and molecular-signalling products in both the publicly funded and private commercial sectors of the biotech industry.

The rapid changes to the very idea of the bio tool that are revolutionising the health sector rely on precisely the two forms of technological intersection described previously: intersecting technological domains—such as molecular marking and the internet—are what enable primer banks and depots to operate at a previously unimaginable scale and speed. But the intersection of technologies with consciousness and worldview matter too: as biotech becomes increasingly focussed on personalised medicine and precision interventions, both the professional scientific communities who develop new 'translational' products and the various constituencies they serve—from patients and clinicians to investors and policymakers—are changing their ideas about what a bio tool can do.

This chapter, which is based on a lecture of the same name prepared for the Intersections conference in Geneva (2014), uses a personal and anecdotal set of examples to track some of the important changes at the intersection of biology, tools, and consciousness that I argue must be understood as fundamental to the rapid process of technological change currently transforming healthcare services in what has been dubbed 'the age of biological control'. Alongside the general questions I am asking here about technological change are some more specific questions about the models of knowledge and uncertainty we use to analyse the process of technological innovation. Personalised and precision medicine, two of the most important new paradigms for understanding a shift away from mass-produced drugs to new bespoke biological products—such as those promised by both the regenerative medicine and the stem cell fields—are often also discussed in relation to preventative and participatory approaches to the management of diseases such as diabetes. 'P4 Medicine', as this approach has been described, offers us a unique opportunity to think about technology as highly intersectional, and in the following examples, my goal is to foreground this intersectionality as an analytic, as well as pragmatic, device.

Cell Networks

It was nearly 9 pm by the time the packed audience of the Wilkins Gustave Tuck Lecture Theatre had decanted itself into the reception room in a far corner of University College London (UCL). Attendance at the meetings of the London Regenerative Medicine Network (LRMN)[2] is always high: it has 6000 members and is the world's largest organisation of its kind. Since 2005, the LRMN has held free monthly meetings enabling scientists, clinicians, entrepreneurs, policymakers, patients, and the general public to attend events that are 'totally focussed on accelerating the attainment of one common goal: the delivery of safe, efficacious therapies that can be affordably manufactured at scale for use in routine clinical practice' (LMRN website). Presentations in this interdisciplinary forum for cellular translation always follow the same format in the tiered, 800-person theatre where they have been held since moving to UCL from Guy's in 2010. The Gustave Tuck theatre is listed on the BBC film-location website, and it readily conveys the sense of a grand occasion. Over 2 hours, three distinguished speakers give 20-minute talks illustrated with lavish PowerPoint slides followed by questions and discussion. Afterwards is a lengthy wine reception where people are encouraged to network. At the reception I ran into a lab director with whom I did fieldwork and confirmed that some of the old stem cell models I used to know had already flown over the rainbow.

'So how are your natural-mutation models doing these days?' I asked the head of the stem cell lab in London where I have conducted fieldwork intermittently for over a decade. 'Nowhere', he said, sounding both defiant and despondent, as he rapidly sank a glass of cold Chablis. 'Why nowhere?' I asked, surprised at his answer. 'No one is interested in human stem cells anymore', he answered. 'Now everyone is using iPS [induced pluripotent stem] cells. They are much better, much faster, they make better models'. So much for the persuasive vision, so recently the subject of such widespread enthusiasm, of human embryonic stem cell models as 'the best tool you can get' for understanding natural mutations: like Dorothy's house in the *Wizard of Oz*, they have been blown away by a tornado of interest in the power of induction. In the fast-paced world of cellular modelling, things had already moved on, again: with a familiar sense of déjà vu, the natural-mutation models derived from donated pre-implantation genetic diagnosis embryos had become models for better models. They were no longer a horizon technology for correcting pathological molecular pathways but instead a sunken sun, the fading rays of which now illuminated a new dish model sparkling with shinier promises.

As my colleague went on to explain, not only had the appeal of iPS cell technology eclipsed its natural progenitor, but hESC cells had become harder to source as well. 'Frozen blastocysts, everyone is freezing their blastocysts. What can you do with a frozen blastocyst? Nothing', he said. The turn in fertility treatment toward single blastocyst transfer, and freezing everything, had vastly reduced the supply of undifferentiated early human material, he lamented.[3] We went on to have a somewhat surreal exchange about the new goal of creating viable human embryos from artificial eggs and sperm. 'Put the progenitor sperm cells in mouse testis you get good sperm', he said. 'Eggs are more difficult'. But haven't they done it in the mouse? 'Yes, in mouse but not in human'. He had got a huge supply of surplus eggs from a US clinic, but he wasn't allowed to bring them into the UK because the US consent process deviated from the Human Fertilisation and Embryology Authority standard. And now that neither eggs nor newly fertilised zygotes were being frozen anymore, it was unlikely more would become available in the future, even if the correct consent procedures could be followed. What about elsewhere, I asked, what about China? You would not want to get eggs from China, he said. Russia? Not from Russia either. Only Europe or the USA. What about Turkey? 'Turkey is going to blastocyst', he replied, shaking his head. 'Everyone is going to blastocyst'. We decided to get some more wine.

Impactful Pipelines

Waiting by the bar amidst the crowded pack of thirsty cell-watchers I reflected on the meeting we had just attended. It was my second meeting of the day, and there had been some unexpected resonances with the first. My first meeting had been an all-day grant holders' workshop at the Wellcome Trust focussed on 'impact'—the irrational obsession of UK higher education for the past ten years or so and the watchword for the all-important Research Excellence Framework, which determines funding levels for all UK universities. Over the course of several talks and informal breakout sessions, a recurring and familiar theme at the Wellcome Trust workshop was the inappropriateness of the impact metaphor and the need for much more dynamic models of how academic knowledge travels and engages its users. One of these alternative metaphors had been 'churn'. 'We have to increase the churn of knowledge', Dr. Jane Tinkler of the London School of Economics and Political Science had counselled, based on her book *The Impact of Social Sciences* (Bastow et al. 2014).

'Knowledge in use' had to be 'churned back' into the existing 'stock' or 'inventory' of 'knowledge not in use', her model suggested, thus 'bringing theory-based knowledge into applied practice'. Social media, Dr. Tinkler suggested, was a key means of achieving this end and thus of getting more 'people value' and 'money value' out of UK social science. The view from the Arts and Humanities Research Council (AHRC) presented later in the day had been slightly different. Dr. Claire Hyland had also used the government's favoured language of pounds and pence in describing the 'paybacks of the creative economy'. Research conducted by the Council had identified benefits in terms of economic capital, human capital, and civic capital, and these could be calculated to show that for every pound invested in the arts and humanities, £10 were generated in the real economy, rising to £20 in the longer term. Whereas the London School of Economics (LSE) speaker had emphasised the mechanism of 'churning' useful knowledge from the academy back into society, the AHRC was focussing on the model of partnerships to promote a model of 'knowledge exchanges' instead of 'knowledge transfer'. The Council's flagship programme of 'sandboxes' linking humanities scholars with partners in business, community organisations, and the general public was proving especially successful, revealing 'new pathways to impact outside of the academy'.

In a concluding panel, the main themes of the day were reiterated: impact is highly varied and diverse, it has variable timespans, and it cannot easily be captured by any metric. Impact measurement has to be highly flexible and plastic. The ubiquitous linear models of pipelines, everyone agreed, were completely inappropriate and misleading. One speaker even suggested they should be banned.

But in the Gustave Tuck theatre later that evening, pipelines to impact were alive and well. As the first speaker, Dr. Natalie Mount, Chief Executive of the London-based promotional organisation Cell Catapult explained, the UK pipelines were not only pumping but were fully primed. Indeed they were full to bursting with 86 novel cell therapies in the final stages of approval. 'There is a big pipeline in the UK which now holds the strongest position in Europe', she confirmed to the audience. Both autologous and allogeneic cell therapies are reaching Stage 3 clinical trials in a wide range of fields, from dentistry and ophthalmology to neurology, oncology, and cardiac care, she noted. New forms of macrophage therapy, adoptive T-cell transfer, and immunotherapy were showing great potential in projects that are the result of significant public investment in this sector—thus confirming the wisdom of the early and substantial UK investment in new cellular therapies. The barriers to translational challenges were being levelled through strategically targeted government-funded initiatives to accelerate the bench to bedside and bench to market the translation of Advanced Therapy Medicine Products, including combination products that integrate cells with medical devices, such as patches and scaffolds.

Pipelines were even more visibly invoked by the second speaker, UCL professor Martin Birchall—Briton of the Year (2009) and leader of the team that developed de-cellularised biologic airway scaffolds combined with autologous cells and stem cells to deliver the world's first successful transplant of a regenerated organ in an adult in 2008 (Macchiarini et al. 2008, p. 2023–2030). In 2010, he performed the world's first combined laryngeal and tracheal transplant with surgeons at the University of California, Davis. After demonstrating that human airway pipelines, like translational ones, are 'more than just tubes', Birchall focussed his audience's attention on the remaining barriers to the eventual 'big breakthrough' and 'big penetration' that would, he predicted, eventually turn regenerative medicine into what he called a 'disruptive technology'—a technological innovation as revolutionary as the Ford Model T he used as his first slide.

Underneath a quotation from Deuteronomy (32:7) inscribed across the mantel of the theatre, Birchall impelled his listeners to imagine the 'big, disruptive leap' that would see conventional organ transplantation replaced by organ regeneration, in the same way horses were replaced by cars. 'In the beginning', he reminded his audience, 'and for a long time after motor vehicles were introduced, horses were still more efficient'.

Birchall's analogy to time, the time of translation we might call it, represents yet another model of 'impact'—and one that reminded me of a workshop I organised on 'The Impact of Impact' in my own department as we all began to struggle with this term. At this workshop, one of my sociology colleagues, Mary Evans, noted 'we are mistaken to believe that the force of an influential idea necessarily comes from the idea itself: the receipt of ideas can be as "active" a force driving their take-up as their production'. This point was echoed by another of my colleagues, Mike Power, who pointed out that for something to have an impact, 'there needs to be a space for that impact to occur. It is a relationship contextualised by many other factors'.

Inductive Reasoning

Back in the Gustave Tuck theatre, a somewhat different version of these observations on the relationalities of impact, or translation, literally hung over the audience's heads. 'Remember the days of old, consider the years of each generation' read the ornate inscription carved into a wooden mantel in large letters across the front of the theatre. It is an odd exhortation from the Song of Moses that continues: 'Ask your father and he will tell you, ask your elders and they will explain to you'. The quotation comes from a famous passage in the Old Testament, when Jacob and his people are led out of the wilderness and instructed in their faith, shown their path, counselled by Jehovah in the laws of righteousness, and reminded of the differences between God and man. Perhaps it was chosen as a metaphor for the passage of knowledge over time, across the generations—and it is undoubtedly an apt phrase with which to crown a University lecture theatre named after an inventor of eyeglasses.

But the relations between students and their elders are often as unruly as those between old and new technologies, as was the case with horses and cars. And this quotation also seemed curiously suited to the apparent discord between the cells and their replacements, the subject of the third and final lecture of the evening. As Professor Philippe Menasche was quick to point out in a surprising opening to the closing lecture, cells themselves may be the biggest disadvantage to the use of new cell therapies. He suggested that it used to be the idea with stem cells that you could generate replacement cells, but attention has shifted increasingly in the direction of potentially repairing or reprogramming existing cells. Building on the model of the paracrine factors involved in cell signalling, Menasche presented a far-ranging vision of induced plasticity delivered through signalling factors extracted from pluripotent cells and repurposed to trigger in situ cellular reorganisation. Such an approach, he argued, would enable the production of more 'cell-friendly' patches for cardiac repair and thus a means of avoiding the 'nightmare' of having to deal with the introduction of actual cells. His vision was of cell therapy without cells.

Menasche's description of the series of translational steps currently needed to successfully introduce allogeneic cell products into the hearts of severely infarcted individuals would have been familiar to most of the audience, as he depicted the standard translational pathway to both cell therapy and tissue and organ regeneration. After selecting the appropriate cell line, it is scaled up through multiple banking stages; then a second selection takes place from this population of cells to identify those that are most suitable for injection into the heart, or into a pericardial patch. This second process of selection is achieved using surface markers and labels that identify cells no longer pluripotent. Also known as 'purification', the second selection process additionally tests for viability and immunogenicity. The cells must be stable and post-pluripotent, as well as safe and quality assured, in order for their 'identity' to be established. And only those with quality-assured identities can be introduced into the patient.

However, even this very rigorous process of removing pluripotency to ensure clinical viability and safety cannot avoid the many risks associated with the introduction of what are, in effect, foreign cells—even if they were originally derived from the patient's own body. The immunological risks of non-endogenous, ex situ, material are considerable—and both unknown and unpredictable reactions comprise a significant threat to the survival of the patient.

Menasche concluded that committed, stable, and high-quality progenitor cells that are sterile, pure, and non-pluripotent can be safely manufactured and supplied, but even after careful profiling and selection, their clinical use poses the risk of a 'nightmare' scenario of cellular chaos no amount of cyclosporine can control.

And so the cells fly over the rainbow once again. Way up high. Indeed where dreams you really dare to dream come true. And skies are blue. As heard in a lullaby ... 'I really believe', Professor Menasche told his audience, 'that what is important is early retention, in order to foster endogenous pathways, which could be feasible if we really target this critical period'. He continued, 'I'm dreaming of cell-free cell therapy. I think cells are simply a nightmare. I think cells are something we had to go through to learn something. But ultimately it is the factors released by the cells that we are after'.

'Remember the days of old, consider the years of each generation' advises the Biblical quotation above his head, in a somewhat oblique reference to memory, genealogy, and duty. The advice seems to apply as much to the cells and cellular processes being described in the lectures as to the lecturers and their audience. Cells remember their origins, much as they can also forget them. They, like Jacob's flock, go to their neighbours for advice and learn from their progenitors. But, like sheep, they do not always do what they are told—and no wonder, since, like people or experimental scientists or students, they undoubtedly receive mixed messages and contradictory advice. They can take unexpected pathways. If we were asked to describe the affective registers of cellular interaction, in either their happiness or their chaos, it is not difficult to imagine them in terms similar to our own—in terms of affection, dislike, happiness, anger, or fear. To 'remember the years of old, and consider the years of each generation' is as easily a parable of pluripotency as it is of righteousness. For indeed, as Professor Menasche pointed out in his highly technical and pragmatic account of cellular-replacement therapy, pluripotency is a very double-edged sword: at once a powerful resource for change and a source of terminal pathology.

A Feeling for Translation

A component of the translational process, as we know, is stabilisation—a close kin to purification or quality control—and a key process in the management of successful pipelines. Stabilisation, like characterisation, takes time—the time of the repeated interactions that become pathways, the time of signalling that becomes development, the time of biological reorganisation as it becomes subject to control, and the time of the repetitive scientific labour required to achieve control over any given system. Stabilisation is thus also the time of communication—including the core communicative interactions through which bench scientists acquire a 'feel' for the system and its components, as well as how those components communicate with each other. This feeling is not only technical or material, in the sense of an acquired ability to handle biological materials with the skill and care necessary for them to be 'happy' or 'stable'. A 'feeling for the organism', as Evelyn Fox Keller described the scientific methodology of geneticist Barbara McClintock, is also an ability to know which components of a system are 'friendly' to one another, to use Professor Menasche's phrase.

The cultivation of friendly feeling returns us to the question of 'partnerships' raised by the AHRC in their account of 'impact', and the 'churn' models proposed by the LSE researchers, reminding us of the many vital partnerships involved in a successful biomedical translation and how they fit into the 'pipeline' model. In asking how we analyse the forms of consciousness and the worlds of value being mobilised in the Gustave Tuck theatre, we might also ask what the social sciences and humanities have to add to the translational process. So often imagined as the bolt-on components to the so-called 'rollout' of bedside translational cellular products, it may be that the humanities and social sciences will need to be integrated much more fully into the 'front' or 'bench end' of the pipeline—but what would such an integration mean in practice? What might a sociologist, for example, contribute to the translational process in order to make it more effective, more 'impactful', as we now say? Or simply more successful?

Already we have some important models of what such a contribution from social science might involve. From Sheila Jasanoff (2004), we have the model of co-production, which recognises that basic scientific research is a thoroughly social activity and that even the most 'technical' interactions between scientists and their research objects are socially mediated (p. 15). 'Does it any longer make sense for those concerned with the study of power to assume that scientific knowledge comes into being independent of political thought and action or that social institutions passively rearrange themselves to meet technology's insistent demands?' she asks (p. 15). This principle has been explored in depth by historians as well, such as Simon Schaffer and Steve Shapin in their highly influential study of Boyle's air pump (1985). This study emphasised, amongst other things, the role of highly structured conversations in the production of successful science as well as a description of scientific truth that remains very influential today—namely, that the collective witnessing of experimental performances is crucial to how their results come to be seen as truthful, valid, and objective.

A social scientific extension of these arguments can be found in Charis Thompson's *Good Science* (2013), in which she argues that the kinds of ethical conversations that have taken place in the context of stem cell research remain an underutilised resource in the production of better science—including better translational outcomes. A more ethical science, she argues, does not need to be simply a more bureaucratic one nor one that merely includes more attention to public engagement. 'After two decades of research and participation in rapidly changing arenas of the biomedical life sciences', Thompson writes, 'I have become a passionate advocate for pursuing simultaneously the best science and the best ethical practice conceived not merely as overcoming ethical barriers to research [but] in relation to a multi-vocal mandate' (p. 64). Thompson describes what she calls 'a flexible architecture of reciprocity' as one of the 'ethical choreographies' informing the multi-faceted public debate about the California Institute for Regenerative Medicine (CIRM), and she argues that despite the many disappointing aspects of this debate (such as its chronic entanglement with the American abortion controversy) and of the CIRM itself (which has not produced any successful translational applications), the recent history of stem cell debate in California has produced important lessons about what she calls 'the curation of relations' amongst donors, voters, scientists, biomaterials, corporate executives, state lawyers, advisory bodies, regulators, and the general public (p. 168).

Thompson's argument is thus one that both emphasises 'churn' and prioritises relationships—including novel partnerships, such as her own role as a sociologist working as a consultant with the new CIRM. Thompson argues that despite the challenges, setbacks, and disappointments, 'the opportunity and challenge for co-produced ethics and science has rarely been greater' (p. 255). Her argument, then, again emphasises the time of translation and the importance of time and timing to translation, including the implications of using social time as a tool in understanding biological processes and bio-politics. The effort to co-produce science and ethics, she argues, need not be seen as a slowing down, because it will ultimately produce more sustainable results if it is built on a stronger, more inclusive, foundation.

So how might we evaluate Thompson's claim that a more robust social dialogue about experimental biomedicine might deliver more sustainable translational outcomes in relation to Professor Menasche's parable of pluripotency described earlier? What kind of 'somewhere over the rainbow' scenario of enhanced social dialogue can speak to the question of how cells interact with each other? How can 'knowledge in use' be 'churned back' into the existing 'stock' or 'inventory' of 'knowledge not in use', thus 'bringing theory-based knowledge into applied practice'—including translational biomedical practice?

It may be easier to answer these questions by examining the past rather than producing aspirational prescriptions for the future, and Charis Thompson's retrospective analysis of the debate over stem cell research in California provides a useful example of such an approach. One of the experiences she describes in her book is the design and delivery of a course on social, ethical, and political issues related to hESC derivation, and six full pages (p. 91–97) list the potential topics for such a course she collected from colleagues, students, and conversations with activists, concerned citizens, and members of the public. The six-page list covers everything from questions about defining life and non-life to do it yourself, and dual-use biology, intellectual property, disability justice, good governance, healthcare costs, and the historical use of science in the state of California. Thompson's book is in many ways a thought experiment about what 'good science' would look like if concerned discussion about more of these issues were to be much more fully integrated into the practice of basic science. Contrary to the knee-jerk response that such an integration is both practically and politically impossible and that it would risk severely restricting, rather than improving, scientific practice, Thompson argues that the potential for win-win improvements to both basic science and the democratisation of science policy have been underestimated.

Translational Gaps

We can extend Thompson's argument even further back if we consider the case of IVF and by asking what is meant, exactly, by translational success. Human IVF is arguably one of the most successful translational biomedical technologies of the twentieth century, and its rapid expansion shows few signs of diminishing. Indeed one of the remarkable things about the history of IVF is not only how quickly its use has been normalised and naturalised, but also how many other technologies its increasingly widespread use has enabled—from intracytoplasmic sperm injection and preimplantation genetic diagnosis to stem cell research, iPS cells, and somatic cell nuclear transfer (Franklin 2013). At the same time, the rapid and largely unregulated expansion of IVF has generated a number of costly outcomes, including a significant increase in the number of multiple births, which bring greatly elevated levels of morbidity and pathology for both mothers and children—the cost of which has been entirely shouldered by overburdened public and private healthcare providers outside the IVF industry. Other costly consequences of IVF include its high failure rate, which causes significant and lasting distress to both patients and the health professionals who may rightly worry that despite their best intentions they have made the lives of those seeking their care worse rather than better. In addition, there are likely to be future costs associated with IVF that remain fully to be characterised—such as the extent to which superovulation may affect women's long-term health and the question of whether culturing embryos generates harmful epigenetic mutations.

A recent proposal from Carl Djerrasi—the Austrian biochemist whose work translating synthetic norestrogens into the contraceptive pill gave rise to another iconic twentieth-century technology—is that IVF both will and should become much more widely used in concert with egg freezing to give women improved choices in balancing work and family life. Coincidentally, Djerassi's proposal (2014) was published in the *New York Review of Books* shortly before the news broke that Apple and Facebook were offering health-insurance coverage for egg freezing. In the future, Djerassi argues, IVF will be mainly used by women who are not infertile, thus greatly increasing its success rates. Moreover, if they have frozen their eggs when they are young and do not have to undergo superovulation immediately prior to embryo transfer, IVF will be much simpler and less unpleasant at age 35 or over. It will thus not only become a less traumatic and more successful procedure, but also a more popular option, and, for a significant sector of the female population in high-level professional jobs, a routine intervention.

The response to the proposals of Djerassi, Apple, and Facebook that the straight to egg freezing and IVF route should be encouraged has of course been very mixed. Is putting maternity on ice not simply postponing the difficulties of combining motherhood with paid employment until a woman is older and has even more professional responsibilities? Does such a solution not preserve or even reinforce the assumption that raising a family necessarily imposes an exclusively female penalty? Why not simply provide more and better child care, paid maternity and paternity leave, and more robust and genuine re-entry support for any parent who has taken time out to look after young children? If such measures have been successfully established in Sweden, France, Iceland, Denmark, and Finland, why not in California?

IVF can be considered a successful technology if it is measured by many standard indicators of translational success, including its rapid expansion, widespread acceptance and use, and financial profitability. However IVF is not the only solution to unwanted childlessness, and there is no doubt IVF's expansion has brought with it some changes that have generated new problems, including more widespread delays in childbearing in many of the world's wealthiest countries. The current form of IVF—with more and more expensive and unproven add-ons such as time-lapse embryo monitoring and chromosomal screening— may not be as effective as newer solutions to infertility, such as low-cost IVF. Successful translation depends in part on the perception of the needs being met. Clinical solutions may not always be available or best-suited to all constituencies. And the very presence of IVF generates new perceived needs, such as the need to cryopreserve ova in order to beat the biological clock.

A short parable might illustrate this point. This is my egg-freezing parable—let's call it 'the farmer's new horse'.

Once upon a time there was a farmer who bought a horse at auction and brought it home. Upon arrival he realised the horse was too big to get through the barn door. He called his farm assistants to help him. 'Put the horse on a diet', one of them said. 'Get the horse a blanket so it can stay outside', said another. An engineer was called. 'Dig a tunnel into the barn', he suggested. The conversation continued late into the evening, and suggestions continued to roll in: 'Take the horse back and buy a smaller horse', 'Teach the horse to hold its breath', 'Build a new barn'. Late at night the farmer's daughter joined in the conversation. 'Why not cut a larger barn door?' she asked. 'Oh, we can't do that', he said. 'Why not?' she asked. 'You know that old barn', the farmer answered. 'We have grown up with it, my father built that door, it has lasted in this family for generations, I couldn't possibly replace it, your suggestion is impossible'.

Somewhere over the rainbow is a larger barn door. And once we get through the door we can see that certain kinds of conversations we did not think were possible might begin to become not only possible but obvious. Instead of a technique-driven holy grail to unlock the potential of cells—potentially by getting rid of cells altogether, after their 'secretomes' have been mapped and extracted, to make more friendly patches, and better translational applications—there might be a research choreography that is more like a conversation. It might help if the conversation included the cells themselves, in a context of more care and attention to what co-production really means. There were good reasons people did not want to get rid of horses in favour of motorcars. In the face of the apparent success of IVF, it behoves us to ask also about the terms through which that success is measured—and indeed even if some of the 'signs of success' might, from another angle, look slightly less successful. When Jacob and his sheep were lost, they depended upon one another to survive. If egg freezing for women is the wrong answer, it is because it is the answer to the wrong question. If we want the right question, we might have to be more inclusive in the process of deciding who gets to be asked what's wrong to begin with.

An obvious problem of the pipeline analogy is its linearity. In reality the translational process is repetitive, reciprocal, and ultimately circular. Over time it proceeds, if it proceeds at all, in fits and starts. Often its main guide is failure. The individuals who lead successful translational efforts cannot simply be technicists, scientists, or professional specialists: they have to be social entrepreneurs, constantly ferrying groups of people and teams to and fro to make the crucial connections necessary to the realisation of their goals. The paediatric teams that make the all-important paediatric plans for young patients undergoing novel and often still experimental medical treatment can't think in pipeline terms. They have to imagine a larger door. They have to fly up high. They can't simply ask technical questions: they have to get a feel for the situation that involves a 'curation of relations' and the skill to open ears and hearts as well as eyes and brains.

And so it may turn out that the 'best tool you can get' is not a cell-based model at all. It may be that the 'best tool you can get' to deliver reliable and efficacious cell-based therapies that will yield important health and wealth benefits, save lives, and relieve human suffering is not an iPS cell but something more like churn or choreography or conversation. We will do well to remember these three Cs in the co-produced future that will involve not only getting a better feel for the iPS cell model, but also greater skill in generating more interdisciplinary conversations. You never know who you will meet on the Yellow Brick Road. It takes more than a wizard to make a new heart.

Conclusion

During the current period of rapid technological change, increasing technological intersectionality, and equally rapid transformations of what is considered to be normal and ordinary, it is essential to think critically about the models we use to both interpret and to plan for new uses for technology. Conventionally, technology is understood to be the application of science: the idiom of translation expresses this relation as both a promissory and a managerial expectation. The translational idiom is promotional and aspirational: it emphasises the goal of delivering the required goods to the right address on time and ready to roll. Impact is a closely related concept that carries with it the implication of failed delivery. Impact is a muscular word derived from physics to describe a collision between two unrelated objects. In the context of research policy in countries such as the UK, impact is intended to demonstrate the government's commitment to make public spending count, with visible and measurable results that can be seen and appreciated by the general public.

The pipeline model corresponds to both of these aspirational policy discourses with a concrete image of 'delivering the goods'. Pipelines provide swift and efficient means to transport goods—be it in the form of the London Tube or your water supply. A direct pipeline is best: it offers the most efficient channel through which to supply a desired good, be it oil or information.

However, an irony of these promotional and aspirational idioms is that they can impede the very flows they are allegedly designed to accelerate. The development of effective technological solutions to problems such as infertility cannot be imagined simply in terms of translational challenges that will become impactful through more efficient pipelines. The future provision of effective cell therapies for myriad health needs undoubtedly requires the step-by-step solution of numerous technical problems by highly skilled clinicians and scientists. But good solutions require a much more circular process as well. Good solutions, for example, require that the right questions are asked to begin with. And these questions have to be asked of a wide range of people. There is, in fact, unlikely to be any one solution that will be right for everyone suffering from the same medical condition. Translation, an idiom borrowed from the humanities, describes the process of *transforming* one language into another. Technological translation is also a process of transformation, and no era before ours has demonstrated how profound and intimate those changes to our world can be. We need better models than pipelines and impact to help us appreciate the complexity of technological change, and above all we need much better conversations to help us guide the future of biomedicine.

Notes

1. See Donald (MacKenzie 1984, p. 473–502) for a discussion of the misuses of this quotation.
2. The LRMN is one of the world's largest independent, non-profit, and community-based networks of cell- and gene-therapy professionals from a wide range of sectors including science, medicine, law, policy, ethics, philanthropy, industry, and government. See more at http://www.lrmn.com/. The session described here took place in September 2014.

3. Blastocysts cannot be used to cultivate immortalized cell populations because by this stage in early embryonic development the cells have begun to differentiate.

References

Bastow, Simon, Patrick Dunleavy, and Jane Tinkler. 2014. *The Impact of the Social Sciences: How Academics and Their Research Make a Difference*. London: Sage.

Djerassi, Carl. 2014. The Divorce of Coitus from Reproduction. *New York Review of Books*, September Issue.

Franklin, Sarah. 2013. *Biological Relatives: IVF, Stem Cells and the Future of Kinship*. Durham, NC: Duke University Press. http://sarahfranklin.com/wp-content/files/The-Impact-of-Impact-Workshop-Report-2010.pdf

Jasanoff, Sheila. 2004. Ordering Knowledge, Ordering Society. In *States of Knowledge: The Co-Production of Science and the Social Order*, ed. S. Jasanoff. London: Routledge.

Landecker, Hannah. 2007. *Culturing Life*. Cambridge, MA: Harvard University Press.

Macchiarini, Paolo, et al. 2008. Clinical Transplantation of a Tissue-Engineered Airway. *The Lancet* 372 (9655): 2023–2030.

MacKenzie, Donald. 1984. Marx and the Machine. *Technology and Culture* 25 (3): 473–502.

Marx, Karl. 1867. *Capital: A Critique of Political Economy*. Moscow: Progress Publishers.

———. 1971. *The Poverty of Philosophy*. New York: International Publishers.

Shapin, Steven, and Simon Schaffer. 1985. *Leviathan and the Air Pump: Hobbes, Boyle and the Experimental Life*. Princeton, NJ: Princeton University Press.

Thompson, Charis. 2013. *Good Science: The Ethical Choreography of Stem Cell Research*. Cambridge, MA: MIT Press.

Wilmut, Ian, Keith Campbell, and Colin Tudge. 2001. *The Second Creation: Dolly and the Age of Biological Control*. Cambridge, MA: Harvard University Press.

Sarah Franklin worked at Lancaster and the London School of Economics before being elected to the Chair of Sociology at Cambridge University in 2011. Since moving to Cambridge, she has secured over £2 million from the European

Research Council, the Economic and Social Research Council, the Wellcome Trust, and the British Academy to establish the Reproductive Sociology Research Group (ReproSoc). Working closely with Nick Hopwood and Martin Johnson, she has also established the IVF Histories and Culture Project at Cambridge—an interdisciplinary initiative exploring both the history of UK in vitro fertilization (IVF) and the culture of mammalian-development biology from which it emerged. Her most recent book is entitled *Biological Relatives: IVF, Stem Cells and the Future of Kinship* (2013).

Intersections of Technological and Regulatory Zones in Regenerative Medicine

Linda F. Hogle

In October 2015, the US White House published a policy document detailing program priorities to develop an American innovation ecosystem entitled 'A Strategy for American Innovation'. Release of the document was timed to coincide with 'Back to the Future Day' (21 October 2015), the day in the Hollywood film in which the main character Marty McFly was propelled 30 years into the future. When the day arrived, popular media reprised the predictions about social and technological changes, showing which ones had been realized and which remained a fantasy. A press release accompanying the document noted that policy set now would set the pace for medical, engineering, and other innovations being imagined another 30 years hence. Buried deep within the document was a call for 'Designing Smart Regulation to Support Emerging Technologies':

> [t]he innovation process is changing. Key trends include the drastic reduction in costs to launch and scale technology; the lack of regulatory

L.F. Hogle (✉)
Department of Medical History & Bioethics, School of Medicine & Public Health, University of Wisconsin–Madison, Madison, WI, USA

© The Author(s) 2018
A. Bharadwaj (ed.), *Global Perspectives on Stem Cell Technologies*,
https://doi.org/10.1007/978-3-319-63787-7_3

pathways for the testing and pilot phases of emerging technologies …
and the shift away from technologies that can be regulated in accordance
with stable categories to technologies that enable and require more fluid
approaches. As the innovation process evolves, the Federal Government
needs to develop new regulatory approaches for both new and existing
regulations… Smart regulation can use cutting-edge technologies to
reduce regulatory burden, aid in regulatory analysis, and better solicit
public engagement in the regulatory process. (White House 2015,
p. 117)

Regenerative medicine features in the document as one of the important biomedical innovations requiring a more fluid, flexible approach.
The field emerged in about the same period of time since young McFly's
leap into the future: the first tissue engineering meetings, sponsored by
the National Science Foundation, were held in 1989, and a decade later,
embryonic and then induced pluripotent stem cell technologies changed
the game, along with innovations in biomaterials science and engineering. In those early years, there were no metrics or conventions for deter-
mining how bioengineered human biological entities might be evaluated
for safety or efficacy. Rather, these novel, hybrid entities were evaluated
by the existing discipline-based communities of practice associated with
the regulatory centers comprising the Food and Drug Administration
(FDA): engineering and materials science experts dominated the FDA
center responsible for devices, chemistry and kinetics for drugs, and biology for biologics and vaccines.

At the time, evidence-based medicine (EBM) was becoming established as an organizing principle for assessing new medical technologies.
With its insistence on systematic, controlled testing and extensive data
collection, large, blinded, randomized, controlled trials (RCTs) became
the method of choice for producing valid evidence of safety and efficacy.
Yet for cell-based products, large-scale trials pose challenges, blinding is
virtually impossible, and endpoints are difficult to establish. All of these
conditions created a dilemma both for scientists and regulators, sparking
a still-ongoing struggle to standardize terminology and definitions, to
determine appropriate testing techniques, and to design metrics to know
when these biological beings—made to do things they would not do in

nature—would be safe for use in humans (Hogle 2009). At the same time, the political and social climate in the US regarding medical innovations was unsettled. A push to accelerate approval of products in the early 1990s (in part by more comprehensive political initiatives to 'downsizing government') was equaled by concerns about both safety and ethical issues regarding the experimental use of novel treatments in very ill patients.

The field is again facing a similar situation of ambiguity and uncertainty. Although expertise at the FDA has matured, the still-unknown complexities of molecular mechanisms in vitro with unpredictable effects in vivo—not to mention the advent of techniques such as gene editing (CRISPR Cas9), novel entities such as tissue organoids, and pragmatic problems inherent in scaling up the manufacture of human cells—challenge the production of meaningful data using existing regulatory analysis and EBM protocols. The push to speed product approvals endures but now in political and economic environments that include differing attitudes toward risk and patients' roles in decision-making. As 'unruly objects' (Haddad et al. 2013), stem cells continue to trouble what have been thought to be 'stable categories' of regulation, in part because of their complexity and the potential for both high risk and high hopes for human health but more specifically, because the ecosystem for institutional practices by which societies can test knowledge claims is equally complex. Such ecosystems are particular to cultural and political moments, colored by historical specificities, and are continually changing.

This chapter situates contemporary debates over regenerative medicine implementation and governance within broader frames of changing ways of thinking about and producing evidence in contemporary biosciences and medicine. Analyses of stem cell and regenerative medicine governance to date have been written within circumscribed technological zones, often limited to what is or is not allowed by regulatory authorities in specific locales, what is or is not an ethical therapeutic application, and the variances across societies. Not only do such discussions ignore the many broader uses of stem cell and related technologies beyond clinical treatments, but more importantly, intersections with economic, political, and other kinds of technological zones are not taken into account. As a result, debates often get stuck in poles represented as patient choice versus paternalistic restriction of unproven treatments, normative practices of knowledge production versus cultural 'ways of knowing', and settled institutions versus resistance and invention.

My interest is instead in understanding how regenerative medicine infrastructures are evolving in interaction with multiple technical, social, and political dimensions of emerging technologies more broadly. Assumptions underlying existing forms of evidence-making in regenerative medicine and other emerging fields are being challenged by new assemblages being formed with the introduction of new tools and techniques for analysis, changing, and often contradictory political-economic exigencies being instantiated into laws, participants taking on new roles, theories about science and governance, and more. As science and technology studies scholars note, scientific knowledge and modes of governing its production are inherently social: the practices of governance are shaped by historical, social, and political influences in interaction with scientific and technological artefacts (Jasanoff 2004). I take up Barry's concept of 'technological zones' to describe the way that social and political domains are not distinct; rather, material artefacts and the social work that surrounds them form collectives that are bounded not by physical territories such as the state or institutions but rather by social practices that enable or constrain their circulation and regulation (2001).

A central thread in the new assemblage is dissatisfaction with existing institutions and the means of producing evidence, as suggested by the policy document that opened this chapter. Disillusionment with EBM as it has been practiced over the past 40 years, combined with the advent of new tools and infrastructures for collecting, producing, and disseminating data, and changing ways of viewing individuals as research subjects and objects, together create the conditions of possibility for alternative ways of producing evidence. It is important to note that material entities as much as concepts, theories, and practices are an active part of the emergence of changing assemblages. So while resistance and countermovements can be influential in governance changes, and debates about the use or exclusion of novel entities based on ethical, safety, or other concerns can stimulate governance actions, such elements rarely act alone. The introduction of tools and 'cutting-edge technologies to reduce regulatory burden and aid in regulatory analysis', as suggested in the White House document, are central: they can both affect and be affected by the way evidence is conceptualized.

For purposes of this chapter, I focus on the incorporation of radically different tools of evidence production and the way they are being incorporated into evolving ways of thinking about legal and regulatory regimes. These examples help to illustrate that what comes to count as appropriate and credible forms of evidence are embedded in social as well as technological practices. Specifically, the shift to data-driven science and the use of computational tools to provide evidence are being built in to the development of concepts of 'smart regulation' and 'learning' healthcare systems that will significantly affect the way therapeutic innovations are adopted and care is delivered. While the zones of regenerative medicine governance span labs, government and company offices, and geographic locales, I focus on the US, where the current historical moment of upheaval in systems for evaluating and paying for healthcare and innovations brings the phenomena of interest into sharp relief.

The chapter proceeds in three parts. First, I review reasons for objections to existing forms of evidence production and briefly explain why EBM is challenging for regenerative medicine. Second, I discuss how, among the technological and social solutions proposed to deal with current limitations, so-called data-driven approaches are being promoted both to produce new forms of evidence and do so more efficiently. These will briefly be discussed, along with their implications. Last, I examine proposed legislation that will have sweeping implications for the regulation and delivery of medical treatments and diagnostics, including stem cells. This and similar legislation in the US and elsewhere incorporate the new data-driven tools and ways of thinking about evidence as part and parcel of new thinking about 'smart' regulation. I begin with a brief review of why evidence is particularly challenging for regenerative medicine.

The Problem of Evidence in Regenerative Medicine

Background: Critiques of EBM

Establishing how best to demonstrate efficacy and safety for novel products is always a challenge. Regulatory standards specify the way data must be collected, organized, and analyzed, but the process of establishing those standards is influenced by the way evidence itself is conceptualized. Which proofs come to count in particular technological zones, produced by whom, and by what means?

The quality and acceptability of evidence produced by different means is often depicted as a hierarchy, with data derived from hypothesis-driven, systematic, experimental research at the top (Knaapen 2014; Lambert 2006). Controlled conditions are meant to distance the investigator from the subject and otherwise reduce bias, and standardized data collection and analysis techniques are meant to enable generalization of findings across populations and settings. Observational research, which includes information from existing sources and observations produced by the patient or their clinician (case studies or narratives from surveys or clinician records, including medical histories), is usually considered to be at the bottom of the hierarchy. Physicians have used observations for many years as they try, adjust, and evaluate experimental treatments on their own patients; however, this creates a greater likelihood of bias and spurious relationships among the data (Benson and Hartz 2000; Young and Karr 2011). More systematic methods, it is argued, not only ensure that effects are consistent but also prevent wasting a significant amount of research funding and resources (Ioannidis 2014).

The randomized clinical trial (RCT) is portrayed as the zenith of the medical research hierarchy. The large-scale, controlled, blinded, prospective RCT has long been viewed as the most valid way of producing data with which to make decisions about medical treatments and thus has become a central component of EBM. RCTs have become so ingrained in medical research that it is difficult to imagine alternative ways of producing evidence of function and efficacy of medical treatments and products. However, there are pragmatic problems that make RCTs increasingly challenging. RCTs are expensive and time-consuming. It is difficult to recruit sufficient participants, especially since most patients do not want to chance being randomized to the control group. Participant noncompliance with the protocol is common, especially when there are side effects, and this can affect validity. Results may not translate to meaningful clinical practice for years, if at all. Clinical trials often fail due to lack of proof of efficacy (51 percent); toxicity (19 percent); or poor recruitment, design, or execution strategy (30 percent). Many commentators cite problems with objectivity, even though blinding and randomizing participants are intended to resolve bias issues (Abraham 2007). Others argue that large-scale RCTs do not add significant information and may not be feasible (Hudis 2015). Furthermore, Begley and Ioannidis (2015) demonstrate that much of current published research is not reproducible, even when submitted to peer review for funding and publication (see also Ioannidis 2014).[1]

Beyond the pragmatic problems, researchers and policy makers alike criticize the assumptions built into the way of thinking about producing evidence. Much of the recent critique of EBM comes from social studies of standardization (Knaapen 2014; Mykhalovskiy and Weir 2004; Timmermans and Berg 2003; Timmermans and Epstein 2010). Often denigrated as 'cookbook medicine', critics denounce the way standardized practice guidelines may affect physician and patient relationships and care choices that both caregivers and receivers might make under specific circumstances (Lambert 2006).

Findings represent an 'average' treatment effect that may not benefit all individuals. The populations in trials are highly selected for eligibility criteria and do not represent broader populations of patients with the condition in which the treatment will ultimately be applied (Angus 2015). Trials often omit patients who may have complex illnesses or co-morbidities or those who may have less access to care and study follow-up, even though these may be more representative of the patients for whom treatment might eventually be used. Critics also worry that quantified outputs (expressed as survival rates, probability of therapeutic effectiveness, or statistical risk) are too often privileged over clinician expertise and judgment and patients' expressed experiences. Standardized practices for an imagined average patient ignore complex cultural conditions and when exported internationally in order to facilitate commensurability fail to take into account local governance practices and priorities (Rosemann 2014). The way evidence is produced may not take into account political and economic realities as well as local cultural issues, and this may affect the way technologies are understood or used.

Evidence-based approaches rely on the presumption that the experimental body can be made into a standardized work object through the use of strict selection criteria and methodological protocols to control as many parameters as possible The assumption that there is a 'universal body' from which knowledge is to be generalized has been challenged, however, first by the recognition that diversity among research subjects was important, making conclusions about whole populations based on the recruitment of homogenous groups problematic (Epstein 2007). This is particularly true as more has become known about genetic variations among individuals as well as the many environmental effects on genetic expression. Current research designs also do not reflect real-world practices or the lived experiences of subjects. Humans are living, interacting beings who are affected by what they ingest or come into contact with and what dietary, rehabilitation, or mental health therapies they use. As Löwy (2003) notes, humans in clinical trials are not passive experimental objects: lived experimental bodies may alter the effects of an intervention as well as interpretation of data produced by the body. There may be unanticipated interactions that may be significant but are not measured by conventional data collection and analysis. Controlled studies using narrowly defined research questions may miss such complexities.

Social scientists have long called attention to these limitations to the production of evidence (Epstein 2007; Marks 2009; Weisz 2005; Wills and Moreiga 2010). Yet it was the rise of interest in personalized medicine that drew attention to adapting trials in a way that recognized individual particularities that might transcend existing disease—or regulatory—categories. Personalized medicine aims to design treatments that consider a patient's genetic, anatomical, and physiological characteristics, which in turn calls for gathering data differently. The Precision Medicine Initiative, for example, argues for real-time collection of data produced by passive and active alternative tracking of information about individuals using devices and other means to produce data that is more patient generated (NRC 2011). The shift to measuring responses in individual bodies rather than populations through trial innovations (such as adaptive and n-of-1 trials) thus went hand in hand with initiatives to conduct 'smart', 'adaptive' evidence production and governance.

Challenges of Proving Safety and Efficacy of Cell-Based Products

Regenerative medicine is inherently a personalized approach, that is, cells and regenerative agents initiate physiological changes in response to the physiological environments into which they are introduced. Many approaches also use a patient's own cells that are treated ex vivo and reintroduced in a sort of auto-transplant. Because of these individualized phenomena, as well other unique properties of cell-based technologies, regenerative medicine faces unique challenges in producing evidence of safety and efficacy. To begin with, pre-clinical tests (animal and lab tests before putting materials into humans) are more challenging than most drugs (Barazzetti et al. 2016). There are no good homologous animal models to mimic human physiology or complex diseases (Perrin 2014). Also, tissue changes from cell treatments may take a long time to appear, but because small animals have a short life span, outcomes may not readily be observed.

There are novel safety issues as well. Unlike most drugs, cell-based treatments cannot simply be stopped if there are problems; cells integrate into tissues, migrate, and can stimulate cascades of signals affecting tissues other than the intended targets. Because they continue to have effects long after a single administration, long-term follow-up is important to know if there are unintended consequences (beneficial or harmful). Imaging and tracking systems with which to see if the cells survive, stay in the intended tissue, or change (e.g., cause tumors or turn into a different tissue than intended) are important but are expensive and unavailable to some researchers. Delivery systems (intravenous injections, cells embedded into a 'patch', or other approaches) may also make a big difference in cell survival and engraftment.

Efficacy is extremely difficult to gauge for cell-based treatments. Tissue naturally remodels over time after injury or intervention, so how would one know if the changes resulted from the stem cells themselves, some effect the cells exerted on endogenous processes (e.g., paracrine signaling), or other native processes completely unrelated to the stem cells? How much cell survival and engraftment is *clinically* significant? Clinical end points are very difficult to determine in cell-based therapies, so clinicians often rely on vague qualitative, observational measures: is a 5 percent increase in ejection fraction significant enough to demonstrate efficacy in a cardiovascular stem cell trial? What about the 'ability to climb a flight of stairs' or 'less pain'? Cardiovascular disease patients notoriously exhibit a placebo effect in about 40 percent of cases no matter what (if any) treatment they receive. How is that to be figured in to evaluations of efficacy? What matters, and how does it matter? As one clinician-researcher put it: 'To be honest, we don't even know if they beat … if function's improved and patients can climb a flight of stairs, who cares?' (Murry as quoted in Couzin and Vogel 2004).[2] Nevertheless, evidence of efficacy given by systematic means has been key to regulatory clearance for use in humans.

To address some of these concerns, recent innovations from computational biology are augmenting older systems of providing proof. For example, human induced pluripotent stem cells are being grown into tissues in vitro to substitute for some preclinical tests. Computational (in silico) methods can then be used to obtain genetic and developmental information to determine whether cells are properly forming and maintaining genetic cues and traits associated with the desired cell types (Mullard 2015).[3] While such means provide a simulation and prediction of probable outcomes, they are not the gold standard of testing in living systems, require significant expertise and access to computing capabilities, and are thus unavailable to researchers without these resources.

Impetus for Change

The difficulties inherent in providing evidence in cell-based technologies, along with disagreements about EBM and complaints about cumbersome and incohesive regulatory oversight, stimulate calls for regulatory change. Conventional methods of providing evidence do not easily fit emerging technologies like regenerative medicine. Consequently, various forms of resistance and workarounds have arisen, particularly in countries outside of more standardized Euro-American regulatory contexts (Bharadwaj and Glasner 2009; Rosemann 2014; Faulkner 2015).

Rosemann and Chaisinthop argue that resistance to EBM is becoming more organized and global in stem cell medicine (2016). They highlight various medical associations that advocate the use of case-based studies and patient self-assessment as a way of producing evidence in regenerative medicine. As a result, there is a pluralization—and perhaps a stratification—of forms of knowledge production in which networks of actors, institutions, and modes of circulating claims work to create an 'alter-standardization' to conventional EBM. The authors conclude that the lines between what is seen as acceptable and unacceptable research practices and forms of evidence production are being negotiated. While many observers would debate the meaning of 'acceptable' evidence, particularly as being used in some of these contexts, the conditions are ripe for such negotiations.

Nonetheless, while there may be cultural, political, and historical specificities engendering some such counter-measures, movements away from EBM to alternative ways of producing evidence in biomedicine are not isolated within particular areas of research and are not just about resistance. In the next section, I provide an illustration of political and social undercurrents that are intertwined with the uptake of new tools of evidence production.

Political Zones of Governance

How Do Governance Systems Learn to Be Smart?

'Evidence-based policy' followed EBM in the 1990s, similarly based on the idea that policy making should be informed by rigorous, testable, quantifiable data based on data rather than ideology or judgments. More recently, however, the term 'smart regulation' is increasingly used, especially in Canada, the UK, and the US, as noted in the opening quote from the White House. Unlike 'evidence-based', 'smart' can mean using better reasoning, being more informed, or simply more efficient, clever, strategic, or more adaptable and capable of learning. In operational terms it could mean using different *kinds* of evidence to inform policy, using less oversight in response to concerns of those being regulated or other ways of 'learning' to govern. For example, in a Canadian policy document similar to the US innovation policy, 'smart' regulation is said to be responsive in terms of giving regulatees the flexibility to determine how results should be achieved as long as 'high standards' are upheld (Expert Committee 2004). A European Commission presentation suggests that smart regulation means policy simplification and bringing in new participants.[4] US policy has recently begun applying such policy orientations to healthcare under the rubric of 'learning healthcare systems' (IOM 2007). This framing of policy reveals much about contemporary political dynamics in countries where these terms are being used. Speed, efficiency, and flexibility are emphasized over cautious control.

Being smarter in the US also entails being more cost efficient, where healthcare reform efforts are attempting to contain costs while transforming the delivery of care. US healthcare is based on a fee-for-service payment system—whereby services are paid for by private insurance companies (hired by individuals or their employers)—or, for elderly and poor patients, a government payer (Centers for Medicare and Medicaid Services, or CMS). Reform legislation passed in 2010 shifts payments to a 'value-based' scheme whereby patient outcomes will be incentivized (payment for quality of care) rather than paying for ordered services (quantity of care).[5] Proving 'value' will require different forms of evidence than that used for EBM clinical effectiveness in that it will incorporate indicators to assess both health outcomes and economic impact.

High-level initiatives leading up to health-reform efforts aimed to align the twin goals of cost efficiencies and personalized medicine. One such effort shows the evolution in thinking about how knowledge should be generated to achieve these aims. In 2006, an advisory body initially called the Roundtable on Evidence-based Medicine was convened by the Institute of Medicine (IOM) with a goal to ensure that 'by 2020, ninety percent of clinical decisions will be supported by accurate, timely, and up-to-date clinical information and will reflect the best available evidence' (IOM 2007). Composed of clinicians, policy makers, and industry representatives, the stated task was to identify barriers to this goal and pose possible solutions through public-private partnerships.

Limitations posed by EBM were identified as key barriers, so the Roundtable set out to re-evaluate standards of evidence. At a workshop entitled 'Evidence-Based Medicine and the Changing Nature of Health Care', health economist Lynn Etheredge advocated for conducting studies in silico rather than using conventional RCTs for many purposes as a way to re-engineer clinical research while lowering regulatory and technical barriers (IOM 2007). Computer-assisted trials and data gathering, he argued, added benefits of speed, comprehensiveness, and low cost, along with the ability to link research and clinical care by continually gathering and analyzing data on all patients in real time. His plan promoted the linkage of data into large, searchable national databases: 'Studies that would now take years will be doable, at low expense, in a matter of weeks, days, or hours' (Etheredge 2007, see also 2014).[6]

Recommendations were thus made to adopt computational methods and data analytics, moving to Bayesian (probabilistic) statistics, simulations, and relational databases rather than relying on canonical causal statistics. The recommendations also emphasized that a re-evaluation of data collection and evaluation more broadly should proceed, with a new infrastructure to incorporate flexibility: *The nation needs a healthcare system that learns* (emphasis added) (IOM 2007, p. 3).[7] Subsequent reports reiterated the theme of learning but added this had to involve continual data collection rather than the conventional method of gathering information at points in time under controlled conditions and continually adapts: 'Improving quality and controlling costs requires moving from [an] unsustainable and flawed organizational arrangement to a system that *gains knowledge from every care delivery experience and is engineered to promote continuous improvement.* In short, the nation needs a healthcare system that learns...' (IOM 2013, p. 135, emphasis added). This data-driven approach represents a very different way of thinking about the collection, analysis, and interpretation of evidence. Collecting data ubiquitously also enables other kinds of work to proceed, including operations and cost accounting.

The emphasis was on speed and adaptability, with such innovations as rapid protocols and adaptive protocols (Montgomery 2016) and n-of-1 protocols (Schork 2015).[8] For regenerative medicine and other 'personalized' medicine treatments, n-of-1 trials will likely be key. In this design, a large amount of data would be collected at more frequent intervals than current studies but on each individual receiving treatment. In contrast to conventional trials, where dosing and timing is planned for large groups, it would be tailored to an individual's metabolic and genomic profile, and changes in dosage or treatment would be tested against that person's own response to other doses rather than to a population. The patient is her own control, and the expense of recruiting and scaling up to a large study is avoided.[9]

The Roundtable was soon renamed the Roundtable on Value & Science-Driven Health Care, signaling a shift from classic EBM to a focus on value that can be generated from using 'smart' tools in the right healthcare classroom.[10] By 2012, the Learning Healthcare System was defined as '… one in which *science, informatics, incentives and culture are aligned* for continuous improvement and innovation … and new knowledge captured as an integral by-product of the care experience' (IOM 2013, p. 136, emphasis added). The vision included real-time access to knowledge, digital capture of the care experience, and incentives aligned for value to encourage continuous improvement, reduce waste, and reward high-value care … and system competencies that refine operations through systems analysis and information development (IOM 2013, Table 5-1, p. 138). The development of data infrastructures and alignment of incentives were seen as crucial to making the envisioned system work (Hogle 2016a).

An expanded report entitled 'Best Care at Lower Cost' made it clear that cost efficiencies have become part of the production of evidence and that this must be done using computational- and systems-engineering approaches (IOM 2013). Reducing transaction costs (e.g., the cost of collecting and exchanging data among clinicians, managers, researchers, or payers) through integrated data collection, it was suggested, would lead to enhanced revenue generation, risk reduction, patient specification, and engagement and other operations efficiencies.

To implement the re-engineered way of thinking about evidence production in clinical research, one report suggested that private insurers should incentivize such changes through reimbursement structures and benefit plans for the insured. As the report states, insurers are not passive receivers of information; rather, they are the arbiters of evidence.[11] Interestingly, whereas insurers typically have a high bar for evidence in order to reimburse for treatments (especially experimental treatments such as stem cells), workshop participant Steven Pearson (an industry representative) expressed a turnabout in the industry's attitude, suggesting the industry would: '...consider an alternative approach to standards of evidence and decision making as we move toward a learning healthcare system: think of a *dial that can move along a spectrum of evidence*—including evidence that is persuasive, promising, or preliminary' (Pearson 2007, p. 172, emphasis added).

Such concepts of 'evolving evidence' or 'coverage with evidence development' have been incorporated into payment policy and presaged new statutory law (see below). CMS, the major public payer in the US, typically has not paid for experimental treatments but has modified policy to allow for emerging technologies for which there may be little evidence or for which evidence may be difficult to obtain. In such cases, the CMS is allowing evidence to be adaptive:

> While CMS has embraced an evidence-based medicine coverage paradigm, CMS is increasingly challenged to respond to requests for coverage of certain items and services when we find that the expectations of interested parties are disproportionate to the existing evidence base. At the same time, we believe that CMS should support evidence development for certain innovative technologies that are likely to show benefit ... but where the available evidence base does not provide a sufficiently persuasive basis for coverage outside the context of a clinical study, which may be the case for new technologies or for existing technologies for which the evidence is incomplete.[12]

These changes appear to open the door for alternatives to evidence production, including observational studies and data-driven, computational methods. Yet relying on observational and computational methods is a reversal from years of EBM. In the next section, I briefly describe techniques involved in data-driven evidence-building to illuminate the stark differences in underlying assumptions about data and how it is constituted.

Data Analytics: New Forms of Reasoning?

Like conventional approaches in EBM, data-driven approaches base decisions on data rather than expert judgment or intuition.[13] Yet while both provide quantified forms of evidence, RCTs ask narrowly defined research questions with a hypothesis to test, use controlled conditions in select groups of subjects, use clinical measurements (such as lab tests, imaging scans, or physiological measures), and are generally prospective. In contrast, data-driven science asks questions in a more open-ended, evolving way, using associative, relational databases to draw conclusions about *patterns* upon which decisions can be made. Data-driven medicine employs computer algorithms to find associations and patterns within data sets rather than using conventional epidemiological statistics to find causal relationships. The introduction of computational tools capable of analyzing human biology at high volumes, faster speeds, and lower cost of computing; algorithms to facilitate analysis of enormous volumes of complex data; and cloud storage to enable the storage and global exchange of such data have created conditions in which data-driven approaches have been rapidly adopted in the biosciences and clinical research (Topol 2011).

There is also increasing interest in combining very large and very heterogenous databases to determine if there are previously unknown associations between phenomena that might not appear using more narrowly defined statistical methods. To accomplish this requires the use of 'Big Data' analytics. The term has been used in various ways, but in general, Big Data is characterized by large volumes, variety (e.g., text-based, numeric, and imaging data), and high velocity (such as real-time data streams from sensing devices) (Mayer-Schoenberger and Cukier 2013).[14] Techniques include machine-learning, an automated way to find patterns in the data while 'learning' to become more accurate and efficient with each iteration and more data. Natural-language processing is also used to find patterns from which to make meanings from text-based data.

The considerable hyperbole about the new approach to research suggests that the new techniques constitute a new way of conducting science. An oft-cited provocation stated that conventional approaches are becoming obsolete with the use of large volumes of data and new methods for correlating findings: 'The new availability of huge amounts of data ... offers a whole new way of understanding the world. *Correlation supersedes causation*, and science can advance even without coherent models, unified theories, or really any mechanistic explanation at all' (Anderson 2008, emphasis added). Another widely cited report claims that Big Data constitutes a 'fourth paradigm' of 'data-intensive' science, posing it as the progressive evolution of previous periods of experimental then theoretical science (characterized by empirical analysis and modeling) followed by computer-assisted simulations (Hey et al. 2009). The current period, the authors claim, is data intensive.

The differences in approach are significant, because most EBM research is consistent with deductive reasoning (i.e., starting with a theory about mechanisms or phenomena and then developing hypotheses that can be tested—sometimes called a 'top-down' approach). Data-driven approaches, in contrast, are more inductive and 'bottom-up' (i.e., there may be no a priori conclusions to test). Inductive reasoning deals with uncertainty and probability and looks for patterns from observations, generating information that can be used to build conceptual models. It is also relational, scalable, and collects extensive rather than representative information. Instead of attempting to eliminate noise in the data, the noise *is* the data.

Kitchin (2014) and Leonelli (2014) are quick to counter that in practice, data-driven science is less inductive than a hybrid of inductive and deductive approaches. Kitchin argues that contemporary techniques blend theoretically informed choices about how best to design research and query the data while leaving insights and conclusions to emerge from the data. It is not, then, an entirely new epistemology of science. Insights and hypotheses are generated from the data, rather than from theory, but are guided by both theoretical and practical knowledge.

In any case, data-driven approaches differ from current dominant evidence-based models in significant ways. The collection of information, study design, and analysis all point to different ways of thinking about evidence as well as what constitutes valid research. Studies can be retrospective, based on mining existing information, or can be adaptive. For example, a growing trend is to conduct retrospective electronic studies by data-mining existing patient records rather than prospective trials (Yamamoto et al. 2012). This is observational, not experimental, data. Using diagnostic codes, laboratory tests, medication histories, and case notes, cohorts of patients with specific criteria can be identified and compared and aggregated with large databases containing millions of case records to increase the power of statistical findings (Kallinikos and Tempini 2014). To illustrate, patients who were given different treatments for a particular diagnosis could be computationally sorted and compared for beneficial or adverse responses without having to recruit them to be in a trial and have an intervention (Elkhenini et al. 2015). This is far less expensive and time-consuming than conventional trials. Significantly, when patients' records from mined databases are used, there is no informed-consent process (assuming the cases are de-identified when extracted). Investigators see this as a way of lowering regulatory burden and speeding the process, plus it removes resistances of patients to being potentially assigned to a control group (Meystre et al. 2008).

Significantly, data-intensive approaches are likely to include information not only produced by conventional measurements but also by patient-generated data, that is, information provided by patients either in narrative form or by using sensing devices or smart phone apps. Recalling that observational data has been viewed as belonging at the bottom of the hierarchy in the era of EBM, this is another reversal.

Advocates of Big Data are quick to note its wide use in consumer and finance industries, and there are intensive efforts to apply it to many areas of healthcare research and care delivery (Krumholz 2014; Roski et al. 2014).

> If you look at other industries, you see there's an evolution over time from expert intuition to simple testing and guidelines to mathematical models

that enable complex learning systems... Medicine is behind the curve—medicine is cautious... We are still in a regime where we are trying to do A vs. B comparisons, but when you have twenty trials comparing A vs. B and they don't agree with each other, that causes confusion. When medicine makes this shift [to learning systems] we will see big increases in quality and efficiency. (Don Morris, quoted in Klein and Hostetter 2013)[15]

While most do not advocate for doing away with RCTs altogether, many see data-driven techniques as an important way to solve the cost and burden problems of RCTs. Some researchers argue for hybrid designs that could function for a variety of purposes: 'a randomized, embedded, multifactorial, adaptive platform (REMAP) trial' could 'incorporate adaptive designs and Big Data to function not just as a research study, but also as a continuous quality improvement program' (Angus 2015, p. 768). Recalling the IOM recommendations to move to adaptive designs and to produce data that can simultaneously help with operational issues such as cost, these approaches become an essential part of the move to learning healthcare systems.

Beyond the scope of this chapter, there are still numerous pragmatic and epistemological problems with big-data approaches (boyd and Crawford 2012). First, there is an assumption of the existence of electronic medical records, systems interoperability, high-throughput computing capability, and other technological and infrastructure issues that may exclude researchers in resource-poor labs or countries from participating in this alternate form of evidence production. There are also concerns about patient privacy with the extensive data collection and sharing these models entail, as well as other ethical concerns (Hogle 2016a). Many scholars are concerned about whether data produced by such means can be clinically meaningful and actionable, as well as the possible misuses of the power of such data collection (Hoffman and Podgurski 2013; Hogle 2016b).

Nevertheless, the stampede of researchers to the new techniques has already begun and in fact is being encouraged by policy in the US and abroad. The final section shows how the new techniques and ways of thinking about evidence and evidence production are now being codified into proposed law.

New Paradigms for Producing Evidence? New Legislation

Recently passed legislation transforms medical innovation processes and reveals much about contemporary political culture in the US. Called the 'Twenty-First Century Cures Act' (PL 114-255 2016), the legislation makes substantial changes to the statutory and regulatory framework of the US Department of Health and Human Services (which governs healthcare delivery) and the FDA (which regulates medical products). The Act has many provisions, including increasing funding for personalized medicine research and regenerative medicine, changing human subjects' protections, and authorizing the broad sharing of personal health information among researchers and other entities.[16] Such provisions are intended to reduce barriers to research and speed the evaluation and approval of medical innovations, and many take up the language from the earlier IOM reports. Significantly, the law makes fundamental changes to what may be considered as acceptable evidence.

In particular, in a section entitled 'Modern Trial Design and Evidence Development', the law instructs the FDA to use observational data in the evaluation of drugs, biologics, and devices. Such data could come from case histories, patient narratives about their own experience with a treatment, data generated from the patient's mobile health devices (smart phone apps, sensors, or telemetry on dedicated devices), clinical-outcome assessments, data-mining of registries or medical records, surveys, insurance/reimbursement claims, and even published journal articles (see Section 3001). Big Data will play a major role in aggregating and analyzing such disparate forms of data. Of particular interest, patient-experience data (e.g., data collected by patients, parents, caregivers, patient advocacy organizations, disease-research foundations, medical researchers, research sponsors, or other parties determined appropriate) can also be used to evaluate product efficacy. Such information is intended to represent the lived experience of patients, which is often lacking in clinical research and is referred to throughout as 'real-world evidence'.[17] Augmenting or substituting alternative, less rigorous forms of evidence shortens the time and cost of studies, lessening the demands on product sponsors. This also upends the hierarchy of evidence, in which observational data has been viewed as less valid than controlled experimental studies.

Additionally, Section 3033 is specific to regenerative medicine products, allowing for accelerated review for 'Regenerative Advanced Therapies' without the convention of risk-based phases of evaluating safety and efficacy. Phase III may be bypassed in some conditions (especially life-threatening conditions) as long as *preliminary* evidence indicates that it has *potential* to address unmet needs. Post-market review can be accomplished by using real-world evidence.

Computational studies may be used along with conventional animal studies and may include in vitro disease-in-a-dish or other cell-based models (Burnstein and Burridge 2014; Mullard 2015). Regenerative medicine thus plays a role in drug studies as well as potential cellular therapies.[18]

The use of surrogate endpoints and biomarkers is also encouraged. The statute states that 'Acceptable biomarkers for supporting investigational use and obtaining drug approval include surrogate endpoints, which may not necessarily reflect or directly correlate with the clinical outcome of interest'. That is, indicators such as the presence of a particular protein, or tumor shrinkage, or a change in a condition such as improved ability to climb stairs would be accepted as evidence of efficacy rather than the actual clinical outcome that is the conventional endpoint (such as improved survival rates). While biomarkers can be useful measures of efficacy in some cases, they may not indicate that the patient's condition has actually improved, and in some cases, focusing on biomarkers has meant that other risks or co-morbidities are missed.[19]

The FDA is further encouraged to use adaptive clinical trial designs and Bayesian probability methods and predictive analytics, as described previously. These are consistent with the expanding uptake of Big Data analytics. Section 3037 of the statute further allows product sponsors to promote the use of off-label healthcare economic information.

More than 1400 lobbying groups actively lobbied the bill, including pharmaceutical and device industries and information technology services industries. This is not surprising, given the number of provisions not only diminishing existing regulation but also favoring the uptake of specific technologies to produce evidence, including genome analytics and in particular information technologies, data management services, and devices, especially tracking devices used to collect observational data (Tahir 2015).[20]

According to some critics, there is reason to be concerned about the close collaborations between the FDA and the industries it regulates in writing the law. For example, the FDA's Center for Devices and Radiological Health and AdvaMed, a medical-device industry trade group, worked together on the proposed language for most of the device provisions in the Act (Perriello 2015; Tozzi 2015). Framed in press releases as patient participation and empowerment legislation, many patient-advocacy groups were also strongly supportive.

While some lauded the increase in funding and accelerated approval of products, others saw the promise of easier access to experimental treatments as sister initiatives to other efforts perceived to provide greater ability to try unapproved treatments. Much of the language lowering the evidentiary bar emphasizes easier and faster access to innovations, echoing language from many 'right-to-try' laws.[21] Some, but not all, language is similar to the failed Reliable & Effective Growth for Regenerative Health Options that Improve Wellness Act, which attempted to make regenerative medicine products more easily available without phase II or III trials, using any type of cell (even though many have not been validated) and with less data. This bill was strongly opposed by the International Society for Stem Cell Research and the Alliance for Regenerative Medicine, with worries about the lack of scientific methods and reliance on qualitative data. The FDA has also expressed a guarded concern about the implications (Marks et al. 2016).

Some commentators note that shortened review times may not improve outcomes; in fact, previous initiatives to speed the review process appear to be associated with increases in morbidity and mortality (Olson 2002). Other critics worry that using less rigorous data may result in faulty conclusions (Avorn and Kesselheim 2015; Wood and Zuckerman 2015). As noted previously, there is a lack of trust in observational studies due to the high potential for spurious relationships in the data, the inability to reproduce findings, and the potential for undue influence in interpretation and publishing results (Young and Karr 2011). Ioannidis (2014) points to the social issues at play: poor-quality peer review in publication and funding, interests of study sponsors in publishing only positive results, corporate influence in the translation of candidate products, and journal and career review and reward systems that favor more spectacular findings over carefully designed studies. Acknowledging that growing use of Big Data and observational data is becoming central to 'learning healthcare systems', Dahabreh and Kent argue for better empirical comparisons between observational studies and RCTs, cautioning that the findings often differ, creating confusing and costly outcomes without an actionable result (2014).

The Cures Act legislation was touted as a symbolic gesture of unity around 'smart' innovation. It also reflects economic and socio-political currents in the US and elsewhere that entail more than the long-standing battles over too-much versus too-little regulation, constant pressures to speed the process of reviewing innovations opposing concerns about safety and 'good science', and rhetoric about access and patient empowerment. At stake are changing notions of what constitutes acceptable evidence as well as the infrastructures being put in to place that not only alter the flow of information and goods but also facilitate other commercial and political interests.

The example of the Twenty-First Century Cures Act illustrates that the forms regulations take are not inevitable; there are political and historical currents within and across locales that affect both the conceptualization and the execution of guidelines and laws. The changing assemblages represented by evolving legislation, the uptake of tools and techniques based on different measurement logics, and political directives by policy and scientific elites set the conditions of possibility for alternative ways of collecting and interpreting evidence for regenerative medicine and other innovations. Yet new infrastructures are built atop old ones, and long-standing normative ways of thinking about research conduct, ethics, and oversight remain. The tensions and negotiations arising within emerging assemblages will be important for analysts of regenerative medicine to follow.

Discussion

In this chapter, I have drawn attention to some of the broader and inter-secting technological zones in which regenerative medicine resides. Resistance to restrictions, ambiguity in regulatory oversight, and disillu-sionment with EBM clearly play a role; however, more fundamentally, there are shifts in thinking about evidence that go far beyond regenerative medicine alone. Stem cells indeed trouble the stable categories set forth by EBM and policy because of their complexity and recalcitrance to exist-ing ways of measuring evidence. The enormous stakes for patients, researchers, and medical industries have as a result stimulated efforts to create alternative pathways, whether or not legitimated by global scien-tific authorities. While a shift to changing techniques may appear to be resistant to EBM, it is not an outright repudiation; rather, new and old concepts of appropriate evidence and how it is acquired may become more hybrid.

More broadly, the confluence of several phenomena is challenging the way knowledge in biomedicine and medical practice is being produced. EBM as a massive, exclusionary standardization project is perceived by some as unsuitable for regenerative medicine. Movements toward patient-generated data and patient entitlements to choose unproven experimen-tal treatments build on the groundswell of activity around personalized medicine and, in some locales such as the US, a privileging of individual autonomy. The push to speed product approvals endures but now in political and economic environments that include differing attitudes toward risk and patients' roles in decision-making. The uptake of new tools and techniques such as Big Data analytics and predictive computa-tion serves economic concerns for systems as a whole well beyond the production of data for specific innovations. Legislative actions built on platforms serving broader political and economic purposes may directly affect the 'institutionalized practices by which members of a society test knowledge claims' (Jasanoff 2004).

One concern in this chapter has been to show how shifts to differ-ent forms of reasoning are taking place as a matter of both pragmatic and epistemological concerns. This is an opportune moment to ask what work we are expecting evidence to do. There are proximate questions, such as what sort of evidence comes to count and why. By what means do societies decide which questions to ask (or not) and which proofs are relevant to those specific questions? Who is participating in decisions about proofs and oversight, in whose name, and for which purposes (whether all participants are at the table or not)? More far reaching are questions about the infrastructures being built to support the new forms of reasoning. If instantiated into policy and practice norms, particular forms of data collection and evidence production enable action across disparate audiences, disciplines, and sets of expertise. With big-data-driven approaches, information collected about individuals can be re-purposed for secondary research, for operational purposes (e.g., cost containment, identification of high-risk or high-cost patients), and for determination of which treatments are 'value based' and worth a society's investment (or private insurer's coverage) (Hogle 2016a, b).

As Bharadwaj suggests, regenerative medicine has become a touchstone for a number of broader social and political phenomena. Stem cells are a medium for exploring ideas about life, knowledge, commerce, governance, and ethics (2012, p. 304). The phenomena I have discussed in this chapter give testimony to this insight. Grounding future research in this awareness will avoid the resort to polarized debates that often fail to understand the complexity and broader implications of struggles over the production of knowledge about emerging technologies.

Notes

1. Evidence as produced by privileged EBM methods has not always been taken up in practice despite its promise of providing a more systematic, scientific basis for clinical decision-making (cf. Lambert 2006). More than a simple matter of behavior change and slow uptake of new ideas, Timmermans and Mauck (2005) position resistance from clinicians within the broader context of actual medical work practices, professionalism expectations, and institutional constraints in which clinicians practice.

2. There has been little guidance on the necessity of knowing specific mechanisms. The International Society for Stem Cell Research Guidelines of 2016 has vague instructions: Section 3.2.3 states that 'complete understanding of the biological mechanisms at work after stem cell transplantation in a preclinical model is not a prerequisite to initiating human experimentation, especially in the case of serious and untreatable diseases' and Section 3.2.3.1 states, 'Before clinical testing, preclinical evidence … should *ideally provide* a mechanism of action … and demonstrate ability to modify disease or injury when applied in suitable animal systems' (ISSCR 2016, emphasis added).

3. Because of the ability to test responses to adventitious agents, drug candidates, and environmental exposures, iPS cells are rapidly becoming a key research tool outside of therapeutic uses (Laustriat et al. 2010). This is likely to become a larger market than therapies.

4. See, for example, Canadian policy at http://publications.gc.ca/collections/Collection/CP22-78-2004E.pdf and the European Commission report at https://www.oecd.org/regreform/policyconference/46528683.pdf

5. The Patient Protection and Affordable Care Act (PL111-148 [2010]) is also known as the ACA or 'Obamacare', after President Obama.

6. As he later put it: 'In the future, biology and medicine will increasingly become 'digital sciences… We need to complete a national system of pre-designed, pre-populated, pre-positioned databases for open science, so researchers can literally log on to the world's evidence base for biomedical and clinical research….'

7. This aim has been restated in subsequent reports: 'Improving quality and controlling costs requires moving from [an] unsustainable and flawed organizational arrangement to a system that gains knowledge from every care-delivery experience and is engineered to promote continuous improvement. In short, the nation needs a healthcare system that learns…' (IOM 2013, p. 135).

8. Adaptive designs use interim findings to alter the course of a trial during its course, which may include modifying randomization (which would change the probability that a patient is assigned to a control or test arm), adjusting patient recruitment, or other decisions that would affect a patient's treatment. These designs have been criticized for eliminating equipoise, diminishing statistical power, and potentially increasing bias, especially the possibility of a Type I error (e.g., rejection of a null hypothesis that is actually true).

9. Such approaches, however, require ready access to relatively sophisticated laboratory tests, including next-generation genome sequencing, which is unlikely to be accessible or affordable to many global sites.

10. The Roundtable has expanded working groups on value incentives, systems engineering, and digital health technology. Members include representatives from the National Institutes of Health, from the pharmaceutical and insurance industries, health economists, and physicians.

11. Insurers, of course, were very supportive of such linkages: data from medical records and claims data linked to clinical trials and registries data could be used to support decisions about which treatments to reimburse and at which rates.

12. Guidance for the public, industry, and CMS staff on Coverage with Evidence Development can be found at https://www.cms.gov/medicare-coverage-database/details/medicare-coverage-document-details.aspx?MCDId=27

13. Although computational tools are thought to distance the 'human' there are nonetheless judgments about what to include or exclude in the making of algorithms that may reflect situated perspectives. There still may be sampling bias and problems with ontology (boyd and Crawford 2012).

14. The term 'Big Data' has been defined as data sets so large and complex as to strain the capacity of conventional information processing and storage technologies. Kitchin (2014) adds that it is scalable and exhaustive.

15. Morris is the scientific officer for Archimedes, a company using simulation models from clinical trials, health records, literature reviews, and more to make predictions about individual patients.

16. In many cases, data sharing will be allowed to proceed without express informed consent. These changes represent a significant departure from current policy, but for purposes of this chapter, the implications are that many uses of information from and about patients would be categorized as minimal risk and could be redefined as 'operational studies' rather than 'research', enabling third parties to access personal health information. Thus, the ubiquitous and continual collection of data ostensibly for research may be used for other purposes, such as cost efficiency or by for-profit entities for commercial purposes. The significant concerns over the collection, surveillance, and use of personal health information are discussed in Hogle (2016b).

17. PL 114-255 114th Congress. Discussing 'valid scientific evidence', Title III Subtitle C Section 3022 505F (b) amends the Food, Drug, and Cosmetic Act, adding the use of 'real-world evidence', defined as 'data

regarding the usage, potential benefits or risks, of a drug derived from sources other than randomized clinical trials'. Available at https://www.gpo.gov/fdsys/pkg/PLAW-114publ255/pdf/PLAW-114publ255.pdf

18. Computational techniques have become increasingly popular as a way to predict cytotoxicity, analyze pharmacodynamics, and more to inform regulatory requirements for risk modeling. More recently, stem cells are being used to screen genetic variants to do drug sensitivity and toxicity testing, disease modeling, and other applications. For example, induced pluripotent stem cells are grown in culture to the stage of small 'organoids' that mimic tissue in the body, then candidate drugs are introduced to test whether there is an adverse or beneficial response rather than administering to a whole organism. This is a sea change in thinking, since conventional experimental science would demand knowing how substances would interact in living, whole organisms rather than project potential responses based on testing in tissue-like composites in the lab.

19. Examples include Avandia, which lowered Hb A1C (an indicator of a patient's blood glucose in the past two months) in diabetic patients but increased the rate of heart attacks.

20. The Healthcare Information and Management Systems Society, a major health information technology advocacy organization, lauds provisions that prevent software from health-tracking devices and smart phone apps from being FDA regulated and argues that eliminating barriers to sharing of personal health information as currently protected by the Health Information Portability and Accountability Act is a leap forward for health-data analysis. The Act also extends brand exclusivity for some drugs, delays entry for generics for others, and allows device makers to obtain third-party assessments if the design or materials used in products changes rather than provide new study data.

21. Although highly controversial, legislation promising accelerated access to unapproved treatments has been enacted in 38 US state legislatures as of this writing, and a federal-level bill has been proposed (Bateman-House et al. 2015; Richardson 2015).

References

Abraham, J. 2007. Drug Trials and Evidence Bases in International Regulatory Context. *BioSocieties* 2: 41–56.

Anderson, C. 2008. The End of Theory? The Data Deluge Makes the Scientific Method Obsolete. *Wired*, June 23.

Angus, D. 2015. Fusing Randomized Trials with Big Data: The Key to Self-learning Health Care Systems? *Journal of the American Medical Association* 314 (8): 767–768.

Avorn, J., and A. Kesselheim. 2015. The 21st Century Cures Act—Will It Take Us Back in Time? *New England Journal of Medicine* 372 (26): 2473–2475.

Barazzetti, G., S.A. Hurst, and A. Mauron. 2016. Adapting Preclinical Benchmarks for First-in-human Trials of Human Embryonic Stem Cell-based Therapies. *Stem Cells Translational Medicine* 5 (8): 1058–1066. doi:10.5966/sctm.2015-0222

Barry, A. 2001. *Political Machines: Governing a Technological Society*. London: Athlone Press.

Bateman-House, A., L. Kimberly, B. Redmond, N. Dubler, and A. Caplan. 2015. Right-to-try Laws: Hope, Hype, and Unintended Consequences. *Annals of Internal Medicine* 163 (10): 796–797.

Begley, C., and J. Ioannidis. 2015. Reproducibility in Science: Improving the Standard for Basic and Preclinical Research. *Circulation Research* 116 (1): 116–126.

Benson, K., and A.J. Hartz. 2000. A Comparison of Observational Studies and Randomized Controlled Trials. *New England Journal of Medicine* 342 (25): 1878–1886.

Bharadwaj, A. 2012. Enculturating Cells: The Anthropology, Substance, and Science of Stem Cells. *Annual Review of Anthropology* 41: 303–317.

Bharadwaj, A., and P. Glasner. 2009. *Local Cells, Global Science: The Rise of Embryonic Stem Cell Research in India*. London: Routledge.

boyd, d., and K. Crawford. 2012. Critical Questions for Big Data. *Information, Communication, & Society* 15 (5): 662–679.

Burnstein, D., and P. Burridge. 2014. Patient-Specific Pluripotent Stem Cells in Doxirubin Cardiotoxicity: A New Window into Personalized Medicine. *Progress in Pediatric Cardiology* 37 (1–2): 23–37.

Couzin, J., and G. Vogel. 2004. Renovating the Heart. *Science* 304 (5668): 184.

Dahabreh, I., and D. Kent. 2014. Can the Learning Healthcare System Be Educated with Observational Data? *Journal of the American Medical Association* 213 (2): 129–130.

Elkhenini, H., K. Davis, N. Stein, J. New, M. Delderfield, M. Gibson, J. Vestbo, A. Woodcock, and N. Bakerly. 2015. Using Electronic Medical Records (EMR) to Conduct Clinical Trials. *BMC Medical Informatics and Decision Making* 15: 8–18.

Epstein, S. 2007. *Inclusion: The Politics of Difference in Medical Research*. Chicago: University of Chicago Press.

Etheredge, L. 2007. A Rapid-Learning Health System. *Health Affairs* 26 (2): 107–113.

———. 2012. http://www.innovationfiles.org/5-qs-on-data-innovation-with-lynn-etheredge/

Etheredge, L.M. 2014. Rapid Learning: A Breakthrough Agenda. *Health Affairs* 33 (7): 1155–1162.

Expert Advisory Committee on Regulation. 2004. *Smart Regulation: A Regulatory Strategy for Canada*. Report to the Government of Canada, Ottawa.

Faulkner, A. 2015. *Special Treatment? Exceptions and Exemptions in the Politics of Regenerative Medicine Gatekeeping in the UK in Global Context*. Working Paper 46, Economic and Social Research Council.

Food and Drug Administration. 2013. *Paving the Way for Personalized Medicine: FDA's Role in the New Era of Product Development*. Report to the Department of Health and Human Services, Rockville, MD.

Haddad, C., H. Chen, and H. Gottweis. 2013. Unruly Objects: Novel Innovation Paths and Their Regulatory Challenge. In *The Global Dynamics of Regenerative Medicine: A Social Science Critique*, ed. A. Webster, 88–117. London: Palgrave Macmillan.

Hey, T., S. Tansley, and K. Tolle. 2009. *The Fourth Paradigm: Data-Intensive Scientific Discovery*. Redmond, WA: Microsoft Research.

Hoffman, S., and A. Podgurski. 2013. The Use and Misuse of Biomedical Data: Is Bigger Really Better? *American Journal of Law & Medicine* 39: 497–538.

Hogle, L.F. 2009. Pragmatic Objectivity and the Standardization of Engineered Tissues. *Social Studies of Science* 39 (5): 717–742.

———. 2016a. The Ethics and Politics of Infrastructures: Creating the Conditions of Possibility for Big Data in Medicine. In *The Ethics of Biomedical Big Data*, ed. B. Mittelstadt and L. Floridi, 397–427. New York: Springer.

———. 2016b. Data-intensive Resourcing in Healthcare. *BioSocieties* 11 (3): 372–393.

Hudis, C. 2015. Big Data: Are Large Prospective Randomized Trials Obsolete in the Future? *Breast* 24 (S1): S15–S18.

Institute of Medicine. 2007. The Learning Healthcare System. In *The Workshop Summary of the Roundtable of Evidence-Based Medicine*, ed. L. Olsen, D. Aisner, and J.M. McGinnis. Washington, DC: National Academies Press.

———. 2013. Best Care at Lower Cost: The Path to Continuously Learning Health Care in America. In *Report from the Committee on the Learning*

Healthcare System, ed. M. Smith, R. Saunders, L. Stuckhardt, and J.M. McGinnis. Washington, DC: National Academies Press.

International Society for Stem Cell Research. 2016. Guidelines for Stem Cell Research and Clinical Translation. http://www.isscr.org/docs/default-source/guidelines/isscr-guidelines-for-stem-cell-research-and-clinical-translation.pdf?sfvrsn=2

Ioannidis, J. 2014. How to Make More Published Research True. *PLoS Medicine* 11 (10): e1001747. doi:10.1371/journal.pmed.1001747

Jasanoff, S. 2004. The Idiom of Co-production. In *States of Knowledge: The Co-production of Science and the Social Order*, ed. S. Jasanoff. New York: Routledge.

Kallinikos, J., and N. Tempini. 2014. Patient Data as Medical Facts: Social Media Practices as a Foundation for Medical Knowledge Creation. *Information Systems Research* 25 (4): 817–833.

Kitchin, R. 2014. Big Data, New Epistemologies, and Paradigm Shifts. *Big Data and Society* 1 (1): 1–12. doi:10.1177/2053951714528481

Klein, S., and M. Hostetter. 2013. In Focus: Learning Healthcare Organizations. *Quality Matters*. Online Newsletter of the Commonwealth Fund. www.commonwealthfund.org/publications/newsletters/quality-matters/2013/august-september/in-focus-learning-health-care-systems

Knaapen, L. 2014. Evidence-based Medicine or Cookbook Medicine? Addressing Concerns over the Standardization of Care. *Sociology Compass* 8 (6): 823–836.

Krumholz, H. 2014. Big Data and New Knowledge in Medicine: The Thinking, Training, and Tools Needed for a Learning Health System. *Health Affairs* 33 (7): 1163–1170.

Lambert, H. 2006. Accounting for EBM: Notions of Evidence in Medicine. *Social Science & Medicine* 62: 2633–2645.

Laustriat, D., J. Gide, and M. Peschanski. 2010. Human Pluripotent Stem Cells in Drug Discovery and Predictive Toxicology. *Biochemical Society Transactions* 38 (4): 1051–1057.

Leonelli, S. (2014) 'What Difference Does Quantity Make? On the Epistemology of Big Data in Biology'. Big Data & Society 1(1): 1–11. doi:10.1177/20539517145432395. Accessed 7 July 2015.

Löwy, I. 2003. Experimental Bodies. In *Companion to Medicine in the Twentieth Century*, ed. R. Cooter and J. Pickstone, 435–450. New York: Routledge.

Marks, H. 2009. What Does Evidence Do? Histories of Therapeutic Research. In *Harmonizing Drugs: Standards in the 20th-Century Pharmaceutical History*, ed. C. Masutti Bonah, A. Rasmussen, and J. Simon, 81–100. Paris: Ed. Glyphe.

Marks, P., C. Witten, and R. Califf. 2016. Clarifying Stem-cell Therapy's Benefits and Risks. *New England Journal of Medicine*, November 30. doi:10.1056/NEJMp1613723.

Mayer-Schoenberger, V., and K. Cukier. 2013. *Big Data: A Revolution that Will Transform How We Live, Think and Work*. London: John Murray.

Meystre, S., G. Savova, K. Kipper-Schuler, and J. Kurdle. 2008. Extracting Information from Textual Documents in the Electronic Health Record: A Review of Recent Research. *Yearbook of Medical Informatics*: 128–144.

Montgomery, C. 2016. From Standardization to Adaptation: Clinical Trials and the Moral Economy of Anticipation. *Science as Culture*. Online in Advance of Publication. doi:10.1080/09505431.2016.1255721.

Mullard, A. 2015. Stem Cell Discovery Platforms Yield First Clinical Candidates. *Nature Reviews Drug Discovery* 14: 589–591.

Mykhalovskiy, E., and L. Weir. 2004. The Problem of Evidence-based Medicine: Directions for Social Science. *Social Science and Medicine* 59 (5): 1059–1069.

National Research Council Committee on a Framework for Developing a New Taxonomy of Disease. 2011. *Toward Precision Medicine: Building a Knowledge Network for Biomedical Research and a New Taxonomy of Disease*. Washington, DC: National Academies Press.

Olson, M. 2002. Pharmaceutical Policy Change and the Safety of New Drugs. *Journal of Law and Economics* 45 (2): 615–642.

Pearson, S. 2007. Standards of Evidence. In *The Learning Healthcare System*, Workshop Summary of the Roundtable on Evidence-Based Medicine, ed. L. Olsen, D. Aisner, and J.M. McGinnis, 171–183. Washington, DC: National Academies Press.

Perriello, B. 2015. FDA, Med Tech in Bed on 21st Century Cures Act. *Mass Device*. Online Newsletter, December 21. http://www.massdevice.com/report-fda-medtech-in-bed-on-21st-century-cures-act/

Perrin, S. 2014. Preclinical Research: Make Mouse Studies Work. *Nature News* 507 (7493): 423–426.

Richardson, E. 2015. Health Policy Brief: Right-to-try Laws. *Health Affairs*. Online, March 5. http://www.healthaffairs.org/healthpolicybriefs/brief.php?brief_id=135

Rosemann, A. 2014. Standardization as Situation-Specific Achievement: Regulatory Diversity and the Production of Value in Intercontinental Collaborations in Stem Cell Medicine. *Social Science & Medicine* 122: 72–80.

Rosemann, A., and N. Chaisinthop. 2016. The Pluralization of the International: Resistance and Alter-Standardization in Regenerative Stem Cell Medicine. *Social Studies of Science* 46 (1): 112–139. doi:10.1016/j.socscimed. 2014.10.018

Roski, J., G.W. Bo-Linn, and T.A. Andrews. 2014. Creating Value in Health Care Through Big Data: Opportunities and Policy Implications. *Health Affairs* 33 (7): 1115–1122.

Schork, N.J. 2015. Personalized Medicine: Time for One-Person Trials. *Nature* 520: 609–611.

Tahir, D. 2015. Interest Groups Seek to Add Goodies to Fast-Moving FDA Overhaul Bill. *Modern Healthcare*, May 12. www.modernhealthcare.com/article/20150512/NEWS/150519957

Timmermans, S., and M. Berg, eds. 2003. *The Gold Standard: The Challenge of Evidence-based Medicine and Standardization in Health Care*. Philadelphia: Temple University Press.

Timmermans, S., and S. Epstein. 2010. A World of Standards But Not a Standard World: Toward a Sociology of Standards and Standardization. *Annual Review of Sociology* 36 (36): 69–89.

Timmermans, S., and A. Mauck. 2005. The Promises and Pitfalls of Evidence-based Medicine. *Health Affairs* 24 (1): 18–28.

Topol, E. 2011. *The Creative Destruction of Medicine: How the Digital Revolution Will Create Better Health Care*. New York: Basic Books.

Tozzi, J. 2015. Safety Outsourced Under Bill Blessed by FDA, Medical Device Makers. *Bloomberg News*, December 21. www.bloomberg.com/news/articles/2015-12-21/medical-device-makers-get-a-little-help-from-the-fda

Webster, A. 2013. Introduction: The Boundaries and Mobilities of Regenerative Medicine. In *The Global Dynamics of Regenerative Medicine: A Social Science Critique*, ed. A. Webster, 1–17. London: Palgrave Macmillan.

Weisz, G. 2005. From Clinical Counting to Evidence-based Medicine. In *Body Counts: Medical Quantification in Historical and Sociological Perspectives*, ed. G. Jorland, A. Opinel, and G. Weisz, 377–393. Montreal: McGill–Queen's University Press.

White House. 2015. *A Strategy for American Innovation*. Report of the National Economic Council and Office of Science and Technology Policy, October. https://www.whitehouse.gov/sites/default/files/strategy_for_american_innovation_october_2015.pdf

Wills, C., and T. Moreiga, eds. 2010. *Medical Proofs, Social Experiments: Clinical Trials in Changing Contexts*. London: Ashgate.

Wood, S., and D. Zuckerman. 2015. The 21st Century Cures Act Could Be a Harmful Step Backwards. *Washington Post*, November 19. https://www.washingtonpost.com/opinions/the-21st-century-cures-act-could-be-a-harmful-step-backward/2015/11/19/919ace5e-8e27-11e5-acff-673ae92ddd2b_story.html

Yamamoto, K., E. Sumi, T. Yamazaki, K. Asai, M. Yamori, S. Teramukai, et al. 2012. A Pragmatic Method for Electronic Medical Record-based Observational Studies: Developing an Electronic Medical Records Retrieval System for Clinical Research. *British Medical Journal* 2: e001622.

Young, S., and A. Karr. 2011. Deming, Data, and Observational Research: A Process Out of Control and Needs Fixing. *Significance*: 116–120.

Linda F. Hogle is Professor of Medical Social Sciences at School of Medicine & Public Health, University of Wisconsin–Madison. She is a Fellow of the Wisconsin Institute for Discovery and member of the BIONanocomposite Tissue Engineering Scaffolds (BIONATES) research group. The research of Hogle includes analyses of social, ethical, and regulatory issues in emerging cell-based and biomedical engineering technologies. In particular, she is interested in the way efforts to standardize emerging technologies affect and are affected by organizational and oversight arrangements. She is researching changing concepts of risk and privacy in data mining and sharing of personal health information. She has served as an advisor to several international research consortia focusing on stem cells and tissue engineering. Her volume *Regenerative Medicine Ethics: Governing Research and Knowledge Practices* (2014) provides practicing scientists and engineers with a more contextualized understanding of the challenges of regenerative medicine governance, especially those less visible in most of the ethics literature on the subject. A recent paper with Klaus Hoeyer, "Informed Consent: The Politics of Intent and Practice in Medical Research Ethics" (2014), critiques assumptions underlying existing informed-consent practices.

Staging Scientific Selves and Pluripotent Cells in South Korea and Japan

Marcie Middlebrooks and Hazuki Shimono

The very public and rather prolonged unraveling of the peer-reviewed and published pluripotent stem cell research claims of South Korean professor, Hwang Woo-suk, in 2005, and the Japanese scientist, Obokata Haruko, in 2014, captured the attention of the South Korean and the Japanese publics, as well those interested in the future of stem cell science and regenerative medicine around the world. Throughout the extensive media coverage of these 'breakthrough' stem cell success stories and the subsequent stem cell research scandals, the public personas of the lead scientists became wedded to the ontological status of pluripotent stem cells in complex and persistently gendered ways. In this chapter, we explore how Obokata's and Hwang's public presences and personal narratives—created in conjuncture with the Korean and Japanese national news media—helped produce international stem cell research results.

M. Middlebrooks (✉)
Ithaca, NY, USA

H. Shimono
Graduate School of Humanities and Sociology, The University of Tokyo, Tokyo, Japan

© The Author(s) 2018
A. Bharadwaj (ed.), *Global Perspectives on Stem Cell Technologies*,
https://doi.org/10.1007/978-3-319-63787-7_4

85

Through our comparative analysis of the Obokata "STAP" (Stimulus-Triggered Acquisition of Pluripotency) stem cell research scandal in Japan and the Hwang Woo-suk cloned (somatic cell nucleus transfer or SCNT) human embryo stem cell research scandal in South Korea, we investigate the intersections between the perceived "integrity" of stem cell science personas and the 'integrate-ability' of these personas into their respective national publics' imaginations.

Along with the growth of 'regenerative medicine', ongoing anxieties about (inter)national economies (Gottweis et al. 2009), and apprehension over declining birth rates, stem cell science technologies have acquired a heightened visibility in South Korea and Japan (Kim, G.B. 2007; Kim, T.H. 2008; Cyranoski 2008a; Gottweis, et al. 2009; Sleeboom-Faulkner 2011; Sleeboom-Faulkner and Hwang 2012). Moreover, as the sites of stem cell science scandals that garnered considerable domestic attention, South Korea and Japan offer interesting vantage points from which a democratic public's engagements with narratives shaping the development of pluripotent stem cell research can be studied.[1]

Public reactions to stem cell science research—as well as the public's perception of the lead stem cell scientists' persona(s) and/or personal presence—may develop in more rapid and archive-able ways in Japan and South Korea, where digital media and internet access is, relatively speaking, widely available. Due, in part, to the high penetration of digital media and concentrated public interest in stem cell research, both Hwang Woo-suk and Obokata Haruko became instantly recognizable public figures even before allegations of ethical misconduct and scientific fraud surfaced. Public responses to the gradual unfolding of accusations of misconduct were frequently voiced online, and both Hwang's and Obokata's autobiographical accounts of their lives and research struggles generated widespread public interest and speculation.

Public interest in Hwang Woo-suk and Obokata Haruko has, to varying degrees, continued. In South Korea, persistent citizen activists and netizens continue to support Hwang and his work through online blogs and communities. Hwang and his new nonprofit, the Sooam Biotech Research Foundation, have continued to appear in international science news with reports of patent applications (Grose 2008; Cyranoski 2008b), more scientific publications (Cyranoski 2007, 2014a), and Hwang's various animal-cloning endeavors (Cyranoski 2007, 2014a; Kim 2009; Oh 2011). In November 2016, a human embryonic stem cell (hESC) line—a line that Hwang asserted was derived from a human blastocyst created via Somatic Cell Nuclear Transfer or SCNT—was officially registered and accepted into South Korea's national stem cell registry after a legal battle. While this now-Korean-government-recognized stem cell line is described as being of unknown origin (Lee 2016),[2] Hwang and his supporters continue to generate and influence public discourse. In Japan, Obokata Haruko has also regained some public support and sympathy with the 2016 publication of her autobiographical STAP scandal account (2016) and subsequent interviews. The question of whether STAP stem cells might yet be reliably produced reportedly remains an interest in some scientific circles (Sato 2015; Goodyear 2016). Moreover, Obokata launched an online 'STAP Hope Page', where she encourages others to attempt to create STAP cells based on the procedures and protocols she supplies.[3] Thus, in both Japan and South Korea, links between the public presence of the stem cell scientist and their promised pluripotent cells continue to resonate with echoes, however faint, of renewed hope and possibility, even long after their stem cell science stories leave the front-page news. Such continued connection between a scientist's ongoing public presence and the latent potential of their ambiguous, openly contested, or even 'scandalous' research results, while not unprecedented, nevertheless, may point to a subtle shift in the public shaping and mass-mediation of science in a twenty-first-century context.

However, before exploring these and other suggestive themes emerging from Hwang's and Obokata's pluripotent stem cell scandals, we must acknowledge some notable contextual differences that influenced the public's response to Obokata's and Hwang's public personas as well as their stem cell research. For example, Hwang Woo-suk's research had, for the most part, introduced the idea of 'pluripotent' and 'patient-matching' stem cells to the general South Korean public in 2004, while Obokata and her STAP stem cells entered the public stage in late 2014—years after Yamanaka Shinya's 2007 success with inducing human somatic (adult) cells into a 'pluripotent' state. Yamanaka's Nobel Prize–winning research made 'pluripotency' a widely recognized word in Japan

(Cyranoski 2015), and significant Japanese state funds had been invested in developing his 'induced pluripotent stem' (iPS) cell technologies that were lauded as being free from the ethical difficulties associated with pluripotent human embryonic cells. Thus, Obokata's STAP research at the RIKEN Center for Developmental Biology (CDB) was evaluated, primarily, within this domestic Japanese research arena where concern for maintaining an international lead in iPS cell research and therapies had prompted a large—some say hasty and lopsided (see Sleeboom-Faulkner 2011; Mikami 2015)—Japanese-state investment. Hence, Obokata's research team and her supervisors at the RIKEN CDB were seen as competing with Yamanaka and his iPS cell research within a domestic Japanese context. In contrast, Hwang Woo-suk's hESC research in South Korea was largely portrayed and publically interpreted as engaging in an international competition or 'race' with high national stakes.[4] Thus, the Hwang scandal prompted a broader and more inclusive sense of national crisis and a more active and organized 'grassroots' or 'populist' domestic public response (Kim et al. 2006; Chekar and Kitzinger 2007; Kim T.H. 2008).

To be sure, the cloned hESC research that Hwang's team attempted represented a more direct challenge to both continental European regulations and the US government's bioethical stance at the time. This is especially clear when compared to Japanese stem cell scientists' alternative approaches like Shinya Yamanaka's iPS cell work and Obokata's STAP research. Hwang Woo-suk's team relied heavily upon women as both paid and unpaid egg donors, and his research raised familiar bioethical discourses about human cloning and the destruction of human embryos and thus triggered arguments about religious difference and cultural relativism that were mixed with an insistent faith in a globalizing biotechnology and 'universal' medical advancements. Ultimately, despite differences between the Hwang scandal and Obokata's fiasco, it is nevertheless clear that the pursuit of pluripotent human stem cells in both Korea and Japan was conditioned by controversy over the use of human embryos as well as regenerative medicine's far-reaching and flexible economic promises of 'embryo-like' pluripotency in an expansive global market. In this sense, both

Hwang's and Obokata's stem cell research activities were clearly shaped, from the start, by the developing global moral economy of hESC science (Salter and Salter 2007) and the promises of pluripotency (Franklin 2005).

The Troubled Category of Personhood

It has long been observed that new technologies carry the potential to facilitate the birth of novel ideas about the inner human psyche (Benjamin 1968) as well as the emergence of externally identifiable 'social types' (Simmel 1969)—that is, those generally recognizable yet uncannily specific types of people that appear as an exactingly individual manifestation of an over-generalized group form. Keeping this in mind, an examination of a couple of well-known biotech personas seems to be, to us at least, a worthy endeavor, especially as various (inter)national audiences have come to anticipate that the nearly simultaneously global (Bauer and Gaskell 2002) news of bioscience breakthroughs is often also a stage for presenting scientist personas. Moreover, we are inspired by Eva Hemmungs Wirten's call for increased attention to 'the troubled category of personhood' (2015, p. 608) in approaching both the history of science and science biographies. Thus emboldened, here, we focus on the 'problematic personhoods' of the once-promising/now-troubled East Asian stem cell scientists whose public personas have been formed and reformed in the spotlight of two high-profile stem cell science scandals.

We follow the 'pluripotent' stem cell scientists—Obokata Haruko and Hwang Woo-suk—as their public personas have been shaped by the shifting place of their stem cell claims in public discourse. We look at the pairing of their public personas with their once promising technologies and consider the public's affective involvement in the changing evaluation of their research. Which signals, signs, or narratives were critical in the public presentation and mass mediation of Obokata's and Hwang's work? Can instructive intersections between Hwang's and Obokata's public personas and their stem cell research stories be found? If so, what can this tell us about the makings of pluripotent stem cells and scientific selves in East Asia?

To address these questions, we consider how national publics come to recognize and understand stem cell science and stem cell scientists, as well as the ways in which particular persons and groups of persons become invested in the personas of scientific figures and thus, perhaps inadvertently, 'personalize' pluripotent stem cell research. To this end, we consider the degrees to which public scientific personas—like that of Hwang Woo-suk and Obokata Haruko—become 'inhabitable'. Here, our interest in the 'inhabitability' of public scientific personas works both with and against a predominately modern and secular ideal of personhood in which individual persons are believed to possess an internal unity and consistency. This ideal of personhood is, perhaps, best captured by the phrase—'being self-possessed' and 'authentic'. However, we recognize that qualities of 'authenticity'—particularly when displayed by public figures that appear before us through digital and other forms of news media—is often a trait a public audience grants or agrees to bestow. Familiar narrative arcs and other recognizable signs may generally enhance a public's recognition of a 'sincere' or 'genuine' story. Such signals also often facilitate the public's ease in accepting certain public personas. It is with these processes in mind that we pose questions about the 'inhabitability' of scientists' public personas. How are certain mandatory markers of personhood, like gender and age, modified and mobilized in a public and potentially controversial field like stem cell science? Can a plurality of persons produce and present pluripotent cells? Or are there certain leanings or even limitations in who 'inhabits' (and thus can mobilize) human agency in science and who represents stem cell science's promises of plasticity?

Below, we present, in turn, key points of interest and some of the particular claims that emerged from Obokata's and Hwang's public presentations of their respective stem cell science projects, along with the domestic news media's framing of and focus on specific details associated with these two science-news pioneers. We suspect that, for the most part, the Japanese and the South Korean publics' understandings of Obokata's and Hwang's varying degrees of 'genuine' personhood significantly influenced subsequent public reactions to their scientific and ethical claims. Although differences may loom large, we are also, nevertheless, struck by the overlapping ways in which both Obokata and Hwang turned to a strong and familial female figure—that of an older maternal caregiver—to anchor both their personas and their stem cell claims in a moral narrative that draws on a simplified 'virtue ethics' that underlines the import of a person's moral character and intentions. We wonder if it is mere coincidence that both Hwang's and Obokata's pluripotent stem cell stories mobilize an appeal to an oversimplified vision of 'virtue ethics' and rely on a foundational 'moral' maternal presence. We speculate that this move, however, subtle or indirect, mobilized certain national-cultural symbols and gendered associations that linked 'pluripotent stem cells' and 'regenerative medicine' with images of intergenerational maternal care. These affective associations and Hwang Woo-suk's scientific persona resonated deeply with South Korean women who mobilized to help Hwang with his research. In contrast, Obokata Haruko's intergenerational 'maternal' garb seems to have provoked suspicion. Simply put, Obokata's scientific persona proved to be less 'inhabitable' than Hwang Woo-suk's. Below, we explore some of the pertinent details that help explain why Hwang's public persona solidified and enhanced his scientific claims, while Obokata Haruko's public persona detracted from her STAP stem cell research claims.

The New Celebrity *Rike Jyo*

'An inspiring accomplishment for a thirty-year-old *rike jyo* [woman scientist]: Obokata Haruko of the Rikagaku Research Institute (RIKEN) in Kobe'.[5] An announcer's voiceover begins the public *NHK* Television's *News Watch 9* story, immediately after a short video clip focuses on Obokata at the press conference where she publicly unveiled her STAP stem cell research results. As you may have guessed, already, a *rike jyo*, or 'woman scientist', is not usually imagined as a conventionally attractive or fashionably dressed person. Yet, as broadcasted in *NHK's* video clips, along with the many other photographs of Obokata that appeared alongside the breakthrough STAP stem cell news, the new *rike jyo* is clearly an attractive young woman with skillfully applied makeup, a well-groomed coiffure, and modest yet fashionable clothing accented with stylish accessories. Particularly, in NHK's coverage, but in other domestic news sources as well, the term *rike jyo* (often written in angular *katakana* script for extra emphasis) borders on becoming a title or byname for Obokata herself. One thing for sure, *NHK* news coverage and the frequent mass-media invocations of *rike jyo* make it clear that Obokata is to become, at least for the time being, the fresh face of Japan's 'woman scientists'.

In Japan, domestic mass-media interest in Obokata Haruko extended beyond mere 'fresh-face' opportunism and embraced the new celebrity *rike jyo's* personal habits and lifestyle choices. Obokata's hobbies, the way she decorated her research team's laboratory, and her pet turtle all became news items deemed worthy of public interest and, later—with the first hints of trouble—public suspicion. Early on, *NHK's News Watch 9* gave Japanese television viewers a tour of the Obokata team's laboratory and supplied plenty of surprised admiration for the pink, yellow, and white pastel colored walls that are also embellished with decals of delightful Moomin (Tove Jansson's imaginary creatures), characters. Even more media attention, however, was focused on the new *rike jyo's* unusual laboratory attire; instead of an ordinary lab coat, Obokata wore a white 'Japanese-style' apron—a 'traditional' Japanese *kappōgi*—while conducting her STAP stem cell research.

While described as a 'traditional' Japanese-style apron, which can be easily worn over a kimono, the *kappōgi* was, in fact, created in the early twentieth century for Japanese culinary-school students (Misaki 2012). Later, as the nation was increasingly mobilized for the war effort, the *kappōgi* became the uniform of voluntary associations of women who helped send soldiers off to war, assisted the families of soldiers, and made arrangements to receive the returning remains of the deceased. After receiving an official mandate from the Japanese military, these voluntary organizations of women—as well as their *kappōgi* uniforms—spread across Japan and, as Misaki Tomeko (2012) notes, the *kappōgi* became a public symbol of wartime maternity—a symbol of the body that regenerates soldiers and the nation and, more widely, a symbol of the 'Japanese mother'. In postwar times, however, the Japanese mass media adopted the image of the 'Western-style' apron as the new symbol of maternal care and 'American-style' consumption and democracy (Misaki 2012; see also Gordon 2012). The wartime militaristic symbolism of the *kappōgi* was largely forgotten, and the 'Japanese-style apron' became simply an old-fashioned and nostalgic symbol of motherhood and was mostly worn by elderly women or 'grandmothers' (*Obāchan*). Thus, to the general public, the *kappōgi* came to invoke and loosely symbolize an older generation of maternal care.

When her STAP stem cell research became big news in Japan, Obokata Haruko recalls briefly explaining to the press during interviews that her now-famous *kappōgi* was a gift from her grandmother (Staff 2014a). The numerous STAP stem cell-related articles published in Japanese broadsheets were filled with references to the new *rike jyo* and her 'Japanese-style apron'. For example, *Yomiuri* newspaper headlines spoke of the '... achievements of the *kappōgi*-wearing *rike jyo*' and the *Nihon Keizai* newspaper announced a new '*rike jyo* pluripotent stem cell revolution' (Tanimoto 2014). Other media outlets reported a sudden increase in white *kappōgi* sales—a putative '*kappōgi* boom' caused by the so-called Obokata effect (Kimura 2014). These reports—like, for example, a *Digital Asahi* newspaper story on the tripling of online *kappōgi* orders [Sato 2014]—emphasized the link between the new *rike jyo* and her white *kappōgi*. Moreover, as part of their early STAP story coverage, *NHK's News Watch 9* even dispatched a reporter to interview elderly women who, it was reported, '...are familiar with *kappōgi*...'.[6] In their news clip, an NHK reporter—equipped with an enlarged picture of Obokata clad in her *kappōgi* and holding a micropipette—asked elderly women for comments. In one televised scene, a woman quickly replies, 'That is the *rike jyo*, right?' Another woman answers with admiration, 'She is really something!' and then, as if playfully prompting a confirmation of men's preferences from the younger male reporter, she adds with a laugh, 'A *kappōgi* is good for women [to wear], no? Don't you agree?'[7] With this mischievous response, the older-woman turns the *NHK* reporter's quick on-the-street inquiry into a question about Japanese men's presumed and underlying preference for young, attractive women who wear 'traditional' Japanese aprons.

As the Japanese news media's interest in Obokata's *kappōgi* persisted, members of the Japanese public wondered exactly whose ideal *rike jyo*

Obokata Haruko was representing as the celebrated new female face of pluripotent stem cell research in Japan. In an article for *Nikkei Business Publications*, Fukumitsu Megumi (2014) gives a detailed point-by-point evaluation of Obokata's public presentation. A self-described middle-aged woman, Fukumitsu invites her readers to consider several more-mature Japanese women's assessments of Obokata's public persona. Having consulted with her female friends, Fukumitsu gives Obokata's performance high points for her 'feminine power' (*joshi ryoku*) as displayed in Obokata's feminine yet vibrantly avant-garde designer ring[8] and Obokata's playful clothing style among other things. Obokata also receives praise for selecting Moomin—rather than Hello, Kitty—decals to decorate her team's colorful lab. However, despite these high marks, Fukumitsu and her friends agree that Obokata's public persona is, nevertheless, ultimately unconvincing. The women's suspicions unanimously coalesce around Obokata's choice of a 'traditional' Japanese-style apron as lab wear. Fukumitsu explains the *kappōgi* mismatch as follows:

> [The *kappōgi*] can couple up with the memory of kind grandmothers or scenes of cooking, so it is useful in representing the warm character of women. However, the maternal associations are too strong so…[a young female *kappōgi* wearer] can be viewed as unscrupulous or shamelessly pandering [to men's longings for 'old-fashioned' women]). At least around me, I have never seen young women wear *kappōgi* except for when they must [serve food] at funerals. So how could [Obokata] wear [a *kappōgi*] confidently in the lab or in front of the media?…Maybe this is the ideal image of a *rike jyo*, but isn't this what *oyaji* [a derogatory term for middle-aged or elderly men] imagine as the ideal *rike jyo*?[9]

As described above Fukumitsu suspects older men's desires have infiltrated and shaped the 'woman-scientist' persona that Obokata presents. More specifically, Fukumitsu reads (hetero)sexual desire in the old-fashioned maternal symbolism of the *kappōgi*. This becomes the reason the *kappōgi* must be rejected by 'respectable' young women who—given their age and gender—scrupulously avoid wearing Japanese-style aprons. In the end, the 'feminine finesse' and stylishness Obokata's public persona displays is betrayed by a 'suspiciously staged' *kappōgi* that, presumably, only appeals to 'stale old men'. Fukumitsu concludes that if Obokata's new *rike jyo* persona had been created by a professional advertising agency, the *kappōgi* wardrobe would be an unforgivably amateurish mistake.

Other reasons—although none as rousing and vivid as Fukumitsu's—were given to heighten the suspicion surrounding Obokata's 'traditional' Japanese-style apron. Journalists (e.g. Hoffman 2014) were quick to point out that a *kappōgi* with its 'wide sleeves' and 'loose fit' was not practical for lab work despite Obokata's claims that her aprons were exceedingly practical and highly functional. As doubts piled up around Obokata's stem cells, Obokata's *kappōgi* seemed to signal and confirm her 'inauthenticity'; her apron became a piece of evidence that 'proved' Obokata's public persona, as well as her STAP stem cells, had been deliberately staged. The public dis-integration of Obokata's mass-mediated persona signaled underlying public suspicions about her 'authentic' personhood and her ability to fully 'inhabit' her 'scientific persona'. Initially, in some Japanese-media discussions,[10] Obokata's youth was interpreted as enabling or facilitating her discovery of a surprisingly simple and straightforward way of producing pluripotent stem cells. Later, however, Obokata's relative youth, along with her gender and habit of wearing a 'Japanese-style' apron instead of a 'Western-style' lab coat, was seen as simply too incongruous to represent a 'genuine' or 'self-possessed' Japanese stem cell scientist.

Throughout the STAP stem cell fiasco, Obokata insisted she had not worn a *kappōgi* to draw more media attention to her scientific work; she consistently maintained that her *kappōgi* was a gift from her grandmother, that she had worn it while performing laboratory work for at least three years prior to her 'breakthrough' STAP stem cell press conference (Staff 2014a; Obokata 2016). Many among the Japanese press and public, however, continued to believe that Obokata's press-conference performance, and *kappōgi*-wearing new *rike jyo* persona had been crafted and staged by others—namely, senior male scientists. Even an independent reform committee, established to propose new measures for preventing further scientific misconduct at RIKEN, concluded in their official report that Obokata's supervisor and STAP paper co-author, Sasai Yoshiki, had orchestrated a 'showy PR strategy' to attract the national media (Cyranoski 2015, p. 602; Lancaster 2016). This independent reform committee—which did not interview Obokata or her supervisor before releasing their official report— appears to have based this conclusion on video footage of 2014 STAP press conferences (Cyranoski 2015). And while built-up frustrations and jealousy over well-funded RIKEN research programs and other institutional inequalities may have fueled the Japanese scientific community's response to the STAP scandal (Cyranoski 2015), Obokata's press-conference performance—particularly her *kappōgi*-clad style—became a highly visible yet circumstantial piece of evidence that 'proved' to many that Obokata's new *rike jyo* persona was fraudulent.

In the wake of the STAP scandal, new government policies for preventing scientific misconduct have been drafted, and the RIKEN CDB has been restructured. Obokata was accused of scientific misconduct and of stealing embryonic stem cells she allegedly intentionally mixed with her STAP stem cell samples (Sato 2015; Staff 2016). Partly in response to these and other allegations, Obokata published her own account of the STAP stem cell scandal in a book-length memoir. In this book, titled *On that day* (*Anohi*), Obokata reiterates her previous claim that she created STAP stem cells and—in addition to recounting her version of various laboratory-related events—Obokata narrates an overarching life story emphasizing the moral intentions and humanitarian inspirations that reportedly fueled her scientific work.

As described in *On that day*, Obokata's determination to study science began in her early childhood with a desire to help others. An elementary-school friendship with a young classmate who developed severe rheumatism at a young age reportedly precipitated Obokata's dream (*yume*) of becoming a scientific researcher. Obokata writes that this early experience of wanting to help her friend 'became a guidepost for my life' (p. 6). Yet, despite her early career successes—including being appointed a research-team leader at RIKEN's CDB—Obokata recalls feeling sadness and frustration when delays and setbacks impeded her stem cell work. During one such noteworthy occasion that Obokata recounts in *On that day*, Obokata's grandmother is credited with reminding her of the ultimate meaning and ethical import of her stem cell research. According to this narrative,

Obokata's grandmother's reassurances and words become another crucial guidepost in her scientific life.

In *On that day*, Obokata recalls finding herself suddenly and unexpectedly on the verge of tears as she tells her grandmother about the difficulties involved in scientific research. In response, Obokata's grandmother comforts her with the advice: 'think about the reason why you do research. Your research is for everyone, right? [Scientific] research is work that benefits everyone' (p. 67–68). Obokata explains:

> Grandmother's words filled my heart. 'I'm glad she is my grandmother', I thought. That evening grandmother was cleaning out her old dresser. She took out a yellowed *kappōgi* and mumbled…'[It was] back when I wore kimonos, so this must be from before the war…' A striped-red *kappōgi* also emerged [from Grandmother's dresser drawers]. 'Grandma, give those to me, please…I'll wear them instead of lab coats'. Grandmother laughed. 'Wear this?' [But] I had decided I'd wear a *kappōgi* so I would remember my grandmother's advice to put my heart [*kokoro*] in the right place and try my best day after day…. (p. 68)

In this manner, Obokata writes about embracing her grandmother's encouraging words and adopting her grandmother's *kappōgi* as a constant reminder of her grandmother's advice. In *On that day*, Obokata repeatedly recollects her grandmother's reassurances and advice, like for example, 'As long as your heart is in the right place, everything will be fine' (p. 164–65). Thus, in Obokata's autobiographical narrative, a grandmother's *kappōgi* becomes a reminder that helps Obokata keep her heart in the 'right place'. Moreover, in this narrative context, the Japanese-style apron serves as a homage to a grandmother who supplies both the emotional support and, perhaps, the intergenerational reassurances necessary for Obokata to continue her stem cell research. And while this narrative of intergenerational inspiration and family ties— symbolized by a *kappōgi*-wearing scientist—may prove to be powerful to many *On that day* readers (the book has become a bestseller in Japan), Obokata's first highly mediated public appearances as the new young 'woman scientist' engendered public suspicions of pluripotent stem cell science scripted by 'old men'.

The 'Charismatic People's Scientist'

'We are simply motivated by the pure desire to alleviate the suffering of people around the world who are afflicted by disease...' (Hwang et al. 2004, p. 156). With these and similar words, South Korean scientist Hwang Woo-suk describes his team's research endeavors as being motivated by global humanitarian goals. Genuine scientific progress, as Hwang explains, demands complete humanitarian dedication such that scientists, who are interested primarily in their own personal gain, inevitably fail, for they lack the fortitude and patience essential for producing great scientific research. Accordingly, Hwang warns young scientists, 'If you just pursue science as a means to gain secular wealth and fame, then you won't be able to endure all the difficulties [of scientific work]' (Hwang et al. 2004, p. 190). Hwang also confidently predicts that only people who possess an enduring and sincere interest in the world can overcome the demands and disappointments of laboratory life (p. 155). In Hwang's well-publicized Korean-language narrative, the great discoveries of science are made by scientists personally inspired by a deep humanitarian drive.[11]

Not surprisingly, in South Korea in 2004 and 2005, the creation of patient-matching hESCs was generally and, for the most part, genuinely believed to be a global humanitarian enterprise. Hwang's public appeal to a simplified and circular 'virtue ethics'—in which moral motivations were almost a prerequisite for creating so-called world-class biotechnologies—reassured and inspired the South Korean public.[12] If, as Hwang explained, great scientific achievements are founded upon sincere human values, then the moral makings of the scientist are crucial to scientific success. While this view of scientific progress is undeniably skewed and selective, Hwang's story—a tale of virtuously motivated hardworking scientists who would transform human lives around the world with a biotechnology of human regeneration—resonated deeply with a beleaguered South Korean public.

Thus, the public emotional appeal of supporting virtuous hardworking stem cell researchers in early twenty-first-century South Korea depended heavily upon Hwang Woo-suk's public narrative and persona. Like Obokata in *On that day*, Hwang speaks of early childhood friendship and his long-standing emotional and moral investments that inspired his ongoing scientific research. Unlike the Japanese new celebrity 'woman scientist', however, Hwang's scientific persona was so convincingly 'inhabitable' that parts of the South Korean public stepped forward to actively connect with and contribute to Hwang's narrative.[13] To understand the cultural politics of these intertwining affective investments, we ask: How did Hwang's public persona engaged the Korean public on such an intimate emotional level.

There are many ways of answering this question, but here we focus on Hwang's autobiographical narrative and the ways in which this narrative helped generate emotional connections and meaningful interchanges between Hwang and parts of the South Korean public. In 2004, a long autobiographical essay that portrays Hwang's life and scientific work in emotionally moving and simple but inspiring language was published— alongside the story of another successful South Korean scientist and the artwork of a prominent South Korean painter—in a large hardcover titled *My stories of life* (*Na ui Saengmyeong Iyagi*). A rich and evocative narrative, Hwang's autobiographical story was a source for many subsequent newspaper articles and news stories as well as a large number of inspiring children's books published in South Korea.[14] Moreover, earlier or later versions of this or similar narratives had circulated through Hwang's own public talks and lectures, as well as in his numerous interviews and televised appearances. As a book, *My stories of life* is beautifully arranged, and Hwang's essay is thematically connected with the words and artwork of the other two contributors. Equally, if not more importantly, the three main contributors[15] are all men born in 1953—the year an armistice agreement stayed the Korean War. In this way, Hwang's extended autobiographical essay appears alongside the stories of now-accomplished men of his same age—men born in a war-torn impoverished land who lived through South Korea's Cold War-era modernization, economic expansion, and subsequent liberal democratization. Thus, *My stories of life* speaks, particularly strongly to and of a South Korean generational cohort who witnessed great social and material transformations. Moreover, this generation—as well as their children and grandchildren— faced continued economic challenges and social uncertainties.

Kim Geun Bae (2007) has explained in detail the process by which Hwang Woo-suk and his stem cell research became a well-publicized part of then-South Korean President Roh Moo-hyun administration's economic policy. Moreover, accounts of the ways in which other influential figures (Cheon 2006) and government R&D policies have shaped the South Korean public's 'sociotechnical imaginaries' (Kim 2014) contribute to our understanding of Hwang's public appeal. Yet the complex affective connections between Hwang and his public remain somewhat obscure. We find a more intimate explanation of the persistent 'inhabitabilty' of Hwang's public scientific persona in South Korea in the emotional performances of Hwang's autobiographical narrative and research stories. Close examination of Hwang's autobiographical narrative(s) as well as some of the public responses to Hwang's public persona and his scientific research story suggest an inspired emotional and virtue-laden exchange unfolding between Hwang and his public that we assert can be understood as enacting a public or 'real' national melodrama of science.

In identifying both Hwang Woo-suk's autobiographical narrative (as well as his public persona) and the strong Korean public support of Hwang's research story as enacting a stem cell 'melodrama', we draw on Peter Brooks' analysis of melodrama as a being much more than a play or performance characterized by emotional excess (1995). He writes, 'Melodrama is…the drama of morality: it strives to find, to articulate, to demonstrate, to "prove" the existence of a moral universe' (p. 20). Hwang's autobiographical narrative and his public story of regenerative bioscience—created with his strong and persistent Korean supporters—seeks to demonstrate, create, and indeed 'prove' a particular—largely nostalgic—'moral universe'.

The 'moral universe' of Hwang Woo-suk's public autobiographical story begins with the emotional ties of the 'traditional' Korean family in a small agrarian village but is based, primarily, on a sacrificial 'motherly love'. In Hwang's narrative, this 'motherly love' is not restricted to humanity or even, necessarily, mothers. Rather, a deep self-sacrificial love characterizes the natural moral relationship of all living beings in Hwang's story. In *My stories of life*, Hwang describes his early relationship with a cow his mother boarded in their poverty-stricken countryside home. In emotionally evocative vignettes, Hwang describes sharing all his childhood thoughts, feelings, and dreams with the cow, and, in turn, it was 'as though [the cow] was confessing her love to me' (p. 41). The sacrifices cows make for human beings—their labor and bodies given for human consumption—impress Hwang, who describes cows as 'sad but beautiful beings' (p. 43).[16] Thus, from the tender age of five, Hwang resolves to devote himself to cows and thus to improving the welfare of his family and that of the farmers in his hometown.

This autobiographical narrative connects with Hwang's first success in attracting national media attention when Hwang announces he has successfully cloned cows.[17] Hwang, as well as the South Korean media, presented Hwang's work with cows as evidence of South Korea's continued scientific and economic advancements while underling connections to a nostalgic Korean past as well as appealing to domestic beef eaters (Kim, G.B. 2007, Kim, T.H. 2008; Hong 2008).[18] Hwang's story of his relationship with cows is also a melodramatic script that supplies an emotional display of the 'virtue' of sacrificial love. In Hwang's autobiography, the warmth of sacrificial love is contrasted with a cold and calculating or 'rational' individual self-interest. This is precisely the kind of cold self-interested behavior that, according to Hwang, will rarely if ever produce significant discoveries in science. Thus, in his narrative, Hwang underscores his own 'scientific virtues' through his emotional devotion to cows when as a young student he refuses—even after being slapped by a teacher—to pursue more promising studies that will lead to future wealth. Instead, Hwang reaffirms his emotional devotion to cows and veterinarian science and thus shows his 'moral capacity' for science. In *My stories of life*, Hwang repeatedly narrates and emotionally 'inhabits'—and thus strives to prove—his moral universe.

In Hwang's public autobiographical narrative—as told in *My stories of life* and many interviews—his mother also plays a key role by embodying the self-sacrificial love that creates his promissory 'moral universe' and the promissory rhetoric of his cloned embryonic stem cell research. Left to raise her children when Hwang's father died, Hwang's mother is described as sacrificing everything for her children and thus hardworking and self-sacrificing 'like a cow' (p. 51).[19] In a passage that draws on a common idea(l) of 'traditional' Korean mothers, Hwang describes his mother and the gift he received from her as follows:

> A big gift I received from my mother is the straightforward honesty of a cow. Mother wasn't always measuring too far in the past and too far into the future and she didn't fuss over the pay or benefit she would receive but

just worked hard with sincerity throughout her life. She never rested—not even one day. From the early hours in the morning, when the stars were still dense in the sky, until late into the night, when the moon was bright, my mother worked tirelessly like a cow for us. Perhaps because I learned by watching my mother…I have just always lived embracing the idea that hardworking focused sincerity, like that of a cow, is the best. (p. 54)

In essence, Hwang's capacity to devote himself springs from his mother's simple self-sacrificing devotion, hard work, and love. The honesty and sincerity Hwang and his mother are depicted as embodying show an intensity and purity of emotion. Such a heightened emotional content and immanent affective capacity point to intense moral imaginings of the past and, by implication, an increasingly precarious present.

The import of this melodramatic narrative and its accompanying moral imaginings become clearer when Hwang describes his research team's endeavors. In *My stories of life*, Hwang sees a sincere devotion to life that is visible in his team's tireless scientific work. He describes the sacrifices of his young researchers who commit themselves to working in the laboratory from early in the morning until late at night and the comradery and the love they share (p. 154). Hwang also articulates his firm belief that scientists who know how to love those living beings around them 'will never become cold scientists who disregard life' (p. 144). Thus, in the story of the Hwang team's research, an intense and (paradoxically) sacrificial 'love of life' simultaneously performs as morality while enabling the great scientific work that is said to be possible only for those with humanitarian inspirations. However, hESC research requires even more. As Hwang explains, his stem cell research team's scientific labor of love is possible only with a complementary and equally sincere gift of love— that expressed by women who donate their eggs with a sincere mind. While expressing his immense gratitude to these donors, Hwang attributes his stem cell research team's success to the 'beautiful and pure hearts' of his egg donors (p. 116).

In fact, the majority of women whose eggs were used in Hwang Woosuk's hESC research received either monetary or other kinds of reimbursement for their donations (Leem and Park 2008; Baylis 2009; Kim Gyeongnye 2010; Tsuge and Hong 2011). Nevertheless, a sizeable portion of women reportedly made voluntary donations of their eggs to Hwang's research (Leem and Park 2008; Tsuge and Hong 2011). Moreover, even after reports of irregularities in Hwang's research protocols and results became public news, a number of women continued to express willingness to donate their eggs (Leem and Park 2008). In *My stories of life*—published after allegations of unethical egg procurement became public but before accusations of data fabrication became widespread—Hwang heavily emphasizes the emotional motivations he describes as underlying voluntary egg donations.

In a section titled 'Beautiful People' (A-reumda-un Saramdeul), Hwang tells stories about people who want to donate their eggs to his hESC research team and thus contribute to his potentially lifesaving research. Hwang describes receiving emails from women who want to donate their eggs after hearing him lecture on his stem cell research. He writes that medical doctors working with severely ill or paralyzed patients often want to donate their own eggs. In particular, the mother with a sick or dying child stands out as a clear example of the beautiful emotions underlying voluntary egg donation. Here again, Hwang contrasts women's self-sacrificial altruistic donations with the presumed self-interests that accompany paid donations (p. 113–15). Hwang writes, 'The meaning of cloning research as a way of helping human life corresponds much better with the meaning of donated eggs....I want to use eggs that have been donated for research for humanity's future rather than eggs that are bought and sold for money' (p. 122). Thus, Hwang's narrative implies that the sincerity or 'purity' of the donor's heart and intentions influences the ultimate meaning—and, we infer, the ultimate 'truth'—of his cloned embryonic research. Such discourse resonates with the personalized emotional power of melodrama, which Brooks characterizes as 'the principal mode for uncovering, demonstrating, and making operative the essential moral universe in a post-sacred era' (p. 15).

As a narrative technique or technology, melodrama has been described as 'radically democratic', and in South Korea, the outpouring of support for Hwang Woo-suk's stem cell research clearly exhibited 'radically democratic' aspects. Particularly after accusations of ethical irregularities in egg procurement emerged, public sentiment and large pro-Hwang demonstrations pushed against government regulations and bureaucratic structures while advocating for—indeed, demanding—the continuation of Hwang's stem cell research. Hwang's public persona and his moral appeal—along with various underground or alternative explanations that circled among the Korean public (Kim Jongmyoung 2007, 2009) as well as shared national interests—inspired strong personal and public responses. Monetary donations were publicly collected and gifts of land announced to assist in the direct public funding of Hwang's science. In later interviews, women described being 'deeply moved' by Hwang's autobiography and thus inspired to donate eggs after reading *My stories of life* (Kim, Gyeongnye 2010, p. 173). In a particularly melodramatic display, 200 of Hwang's female supporters gathered outside his university office in 2005 to show their willingness to donate eggs for Hwang's research. The women—many of whom were members of the I Love Hwang Woo-suk internet café club—came bearing bouquets and flooded Hwang's office with flowers. The event drew on nationalist sentiment with clear national symbolism but in a particularly melodramatic way. The personal names of would-be egg donors written on individual flower petals seem to epitomize a notable feature of modern melodrama. As Brooks explains it, modern melodrama seeks transcendent meaning but in highly personalized and intimate emotional terms.[20] The giving of eggs to Hwang Woo-suk becomes one such intimate emotional and moral act that promises to connect women donors to a larger national and global 'transcendent' bioethical life. In this and other ways, Hwang Woo-suk's melodramatic autobiographical narrative and his public scientific persona proved to be 'inhabitable' to members of the South Korean public and even moved some women to join in with Hwang's stem cell science drama.

Conclusion

The most immediately recognizable and, perhaps, the most mundane similarities seen in Obokata Haruko's and Hwang Woo-suk's public scientific narratives—as well as their public scientific profiles—are their descriptions of their lifetime dedication to their scientific goals. Both Hwang and Obokata emphasize the long hours they and other researchers spend working. Moreover, both the Japanese and South Korean mass medias, respectively, highlight Obokata's and Hwang's tireless efforts and long research hours. Early *NHK* Japanese-language coverage of Obokata's STAP stem cell news includes video segments of Obokata recounting the numerous times she worked by herself late into the night. 'There were so many days I just wanted to give up and so many nights I stayed up crying…' she says in one popular video clip.[21] Obokata's friends and acquaintances describe her as energetic and motivated despite spending long days and nights busy with her research.[22] Similarly, in South Korea, it was widely reported that Hwang Woo-suk slept just four hours a night and led his team on a grueling schedule that began at six in the morning and ended late at night without any weekend or holiday breaks (Hwang et al. 2004). In fact, early domestic-media coverage of Hwang and Obotaka emphasized uplifting and inspiring elements of their reported achievements.

However, as the South Korean and Japanese 'breakthrough' stem cell research stories continued and questions arose about research irregularities, the scientific persona of Hwang Woo-suk proved to be much more 'authentic' or 'inhabitable' for the South Korean public than Obokata's public persona—as the new 'woman scientist' or *rike jyo*—was to a domestic audience in Japan. As discussed previously, Hwang Woo-suk's melodramatic autobiographical narrative and his public scientific persona enacted emotional and moral 'truths' that resonated with the South Korean public. Moreover, the increasingly precarious position of many South Koreans made various members of the Korean public particularly receptive to melodramatic accounts. Thus, when questions were raised about the Hwang team's stem cell research results, Hwang's public reassurances and his 'sincere' and 'moral' public persona enacted his hESCs claims. Public participation in Hwang's science story—be it the pledges of egg donors, the collection of money to fund Hwang's research, or support expressed through political activities and direct public protests—began to perform or enact Hwang's personal narrative of science. In contrast, Obokata Haruko's new *rike jyo* persona was so 'uninhabitable' and quick to draw suspicion that when questions were raised about her STAP stem cells, the 'authenticity' of Obokata's presence and her self-presentation seemed to quickly dissolve. Despite her decision to use her grandmother's 'traditional' Japanese apron when in the lab, Obokata became a figure who, some believed, was dressed by her laboratory superiors. Thus,

Obokata could not wear her grandmother's clothes as her own, so to speak. Hwang, in contrast, convincingly adapted his mother's 'self-sacrificing love' into a melodramatic narrative and hence invited others to enact this love and play a part in South Korea's stem cell science story. Both Obokata Haruko in *On that day* and Hwang Woo-suk in *My stories of life* used their autobiographies as a way to generate and regenerate their scientific selves and their stem cell claims. In the end, Obokata was stripped of her doctoral degree and no longer conducts research, while Hwang Woo-suk heads the Sooam Biotech Research Foundation, a non-profit organization, established with the help of his supporters in 2006.

For now, Hwang's most public melodramatic narrative—which began with the emotional and material exchanges between cows and humans—has moved on to mine the emotional (and economic) ties between dogs and human beings. On its website, the Sooam Biotech Research Foundation offers to clone any dog regardless of its size, age, or breed and heal broken hearts.[23] The personal emotional narrative Hwang effectively developed to explain his hESC science to a South Korean public has been translated into the science of canine cloning. The sincere devotion of Hwang and his Sooam research team—as well as the healing and regenerative promise of Hwang's cloning science—has been narrated by Peter Onruang, one of Hwang's customers:

> Dr. Hwang and his staff welcomed me with open arms. It was as though I had been away on a long trip and had finally come back home. My first time meeting with Dr. Hwang was very heartfelt and warm. I sat with him for a moment and showed him photos of Wolfie and Bubble. His eyes looked carefully at images of my two dogs. Then he looked up at me and told me that the Wolfie clones should not grow up alone. He was going to clone Bubble for free![24]

Onruang—who maintains a website that encourages others to consider cloning their dogs at Sooam—explains that his dog, Wolfie, changed everything by bringing love into his life. Heartbroken when Wolfie died, Onruang describes Hwang Woo-suk as 'the man responsible for mending many hearts'.[25] This and other personalizeds narratives in which Hwang Woo-suk and his biotechnology are a visible and affecting or emotionally moving public presence, can be compared with the turtle which appears in some early news reports about Obokata's STAP stem cell research in Japan. News photographs show Obokata feeding her pet turtle in the laboratory, and when asked in press interviews what she does in her spare time, Obokata explains that sometimes she talks to her turtle, among other things.[26] Thus, Obokata's pet turtle remains largely disconnected from any overarching story of her science in contrast to the cows and dogs that carry the veterinarian Hwang Woo-suk's public narrative of life science and stem cell science melodrama forward.

Notes

1. See Sleeboom-Faulkner and Hwang (2012) for an examination of some of the differences between Japan's and South Korea's human embryonic stem cell research (hESR)–related decision-making processes and differences in the degree of the national public's participation in the discussion of hESR-related bioethical issues, among other things.
2. An investigative committee at Seoul National University, where Hwang Woo-suk was employed when this embryonic stem cell line (NT-1) was created, concluded in their report that the embryonic stem cells called NT-1 had been derived from a cell that underwent parthenogenesis rather than somatic cell nuclear transfer (SCNT).
3. See https://stap-hope-page.com/. Accessed October 2016.
4. Although Hwang clearly did face some internal domestic competition, he was receiving, by far, the most government funding and support. See Kim (2011) for an account of how this high level of government funding may have increased Hwang's perception of the competitive pressures he faced within South Korea.
5. https://www.youtube.com/watch?v=iPg99dyp694&feature=youtu.be (0:16). First broadcast January 2014. Accessed 25 May 2016.
6. https://www.youtube.com/watch?v=iPg99dyp694&feature=youtu.be (2:46). First broadcast January 2014. Accessed 25 May 2016.
7. https://www.youtube.com/watch?v=iPg99dyp694&feature=youtu.be (3:04). First broadcast January 2014. Accessed 25 May 2016.
8. The designer here is Vivienne West, remembered for her early punk style. Fukumitsu (2014) notes that Vivienne West was popular in 1980s Japan but has more recently come to signify something more unique and edgy than the more commonplace uniqlo styles.

9. http://www.nikkeibp.co.jp/article/column/20140327/390016/?rt=nocnt. Accessed 25 May 2016.

10. See conversation between *News Watch 9*'s Kensuke Okoshi and science expert Mushiaki Hideki. https://www.youtube.com/watch?v=iPg99dyp 694&feature=youtu.be (12:28–12:53). Accessed 25 May 2016.

11. Hwang writes, 'I believe that there are hardly any scientists who conduct research only because of their own curiosity or for economic profit. Such motivations would make it difficult to sustain the extreme patience that the [scientific] discovery process demands. The impetus or inspiration which helps scientists continue on despite great disappointment is an abiding love and interest in people and the world they live in' (Hwang et al. 2004, p. 155). See also Hwang et al. (2004, p. 33) for a similar argument.

12. We readily acknowledge a number of factors—of significance, South Korean expectations of gaining economic wealth and financial stability from stem cell technologies (see Kim Geun Bae 2007 or Kim Tae-Ho 2008 for a more extensive discussion of these factors). Nevertheless, it is important to remember that Hwang's stem cell research was publicly imagined as benefiting both 'global humanity' and the 'Korean nation'. In this context, 'nationalism' or 'national interests', which are too often automatically associated with an exclusionary provincialism, become a form of 'universalism'. Thus, for these and other reasons, we (like Leo Kim 2011, p. 215, and Sang-Hyun Kim 2014, p. 300) disagree with Gottweis and Kim's (2010) characterization of the South Korean public's support for Hwang and his human embryonic stem cell research as 'bio-nationalism'.

13. It is important to remember that Obokata's own personal narrative as it appears in *On that day* emerged only after the STAP scandal had attracted much attention and approximately two years after the first 2014 press conference that introduced Obokata to the Japanese public. In contrast, Hwang Woo-suk had appeared in national (and international) news many years prior to the eruption of his stem cell research-related scandal. Thus, elements of Hwang's narrative and life story had already been presented to the South Korean press long before the publication of Hwang's influential autobiographical story.

14. In 2005, at least 12 Korean-language books for children and young-adult readers were published about Hwang Woo-suk's life story (Cheon 2006, p. 412).

15. In addition to Hwang Woo-suk, Choi Jaecheon, an evolutionary biologist who studied with E.O. Wilson, and the painter Kim Byungjong, appears in *My Stories of Life*.
16. Hwang concludes that cows are 'sad but beautiful beings' (p. 63–64).
17. In 1999, Hwang Woo-suk announced the birth of a 'cloned' dairy cow named *Splendor* by the head of Korea's Ministry of Science and Technology at Hwang's request. Approximately a month later, news of another cloned cow named *Hwang Jin-I* by then-Korean President Kim Dae-jung was announced (Kim Geun Bae 2007).
18. Kim Geun Bae (2007) particularly focuses on how Hwang's work with the artificial insemination and the cloning of cows laid the foundations for his later work with cloning human embryos and stem cell research.
19. 'My mother—in sacrificing everything for her children—was like a cow [sacrificing all]' (p. 51).
20. Consider, for instance, the half-humorous words of a male Hwang supporter in his thirties: 'I have never loved anyone like Dr. Hwang. I spent tremendous time in Hwang-supporting activities. If I had invested so much effort like this, I could have had a British woman from a very wealthy family [laugh]'. See Kim Jongyoung (2009, p. 679).
21. https://www.youtube.com/watch?v=iPg99dyp694&feature=youtu.be (0:16). First broadcast January 2014. Accessed 25 May 2016.
22. A Japanese co-worker who Obokata worked with at Harvard University describes Obokata as a 'gambaruya-san'. https://www.youtube.com/watch?v=iPg99dyp694&feature=youtu.be. First broadcast January 2014.
23. See http://en.sooam.com/dogcn/sub01.html. Accessed December 2016.
24. See http://www.myfriendagain.com/dog_cloning_story.html. Accessed 25 May 2016.
25. Ibid.
26. See https://www.youtube.com/watch?v=92RHJ6RStfE

Bibliography

Abelmann, Nancy. 2005. Melodramatic Texts and Contexts: Women's Lives, Movies, and Men. In *South Korean Golden Age Melodrama: Gender, Genre, and National Cinema*, ed. Kathleen McHugh and Nancy Abelmann, 43–64. Detroit, MI: Wayne State University Press.

Bauer, M.W., and G. Gaskell. 2002. *Biotechnology the Making of a Global Controversy*. Cambridge, UK: Cambridge University Press.

Baylis, Fracoise. 2009. For Love or Money? The Saga of Korean Women Who Provided Eggs for Embryonic Stem Cell Research. *Theoretical Medicine and Bioethics* 30: 385–396.

Benjamin, Walter. 1968. The Work of Art in an Age of Mechanical Reproduction. In *Illuminations*, ed. Hannah Arendt, 217–252. New York: Schocken Books.

Brooks, Peter. 1976. *The Melodramatic Imagination: Balzac, Henry James, Melodrama, and the Mode of Excess*. New Haven, CT: Yale University Press.

Chekar, Choon Key, and Jenny Kitzinger. 2007. Science, Patriotism, and Discourses of Nation and Culture: Reflections on the South Korean Stem Cell Breakthroughs and Scandals. *New Genetics and Society* 26 (3): 289–307.

Cheon, Jeonghwan. 2006. Hwang Woo-suk satae' ui daejung hyeonsang gwa minjokju-ui'. *Yeoksa Bipyeong* 77: 387–413.

Cyranoski, David. 2007. Disgraced Cloner Woo Suk Hwang Attempts a Comeback. *Nature*, December 21. nature.com/news/2007/071221/full/news.2007.400.html. Accessed 10 June 2016.

———. 2008a. Stem Cells: A National Project. *Nature News* 45 (17): 229.

———. 2008b. Hwang Work Granted Patent. *Nature News* 2: 571.

———. 2014a. Cloning Comeback. *Nature News* 505 (23): 468–471.

———. 2014b. Cell-Induced Stress. *Nature News* 511 (July 10): 140–143.

———. 2015. Collateral Damage: How a Case of Misconduct Brought a Leading Japanese Biology Institute to Its Knees. *Nature News* 520: 600–603.

Franklin, Sarah. 2005. Stem Cells R Us: Emergent Life Forms and the Global Biological. In *Global Assemblages: Technology, Politics, and Ethics as Anthropological Problems*, ed. Aihwa Ong and Stephen Collier, 59–78. Oxford: Blackwell Publishing.

Fukumitsu, Megumi. 2014. Mou hitotsu no Obokata mondai o katte ni saiten Vivienne Westwood, Moomin, and Kappōgi. *NikkeiBiz*, March 28. nikkeibp.co.jp/article/column/20140327/390016/?rt=nocnt. Accessed 26 May 2016.

Geesink, I., B. Prainsack, and S. Franklin. 2008. Guest Editorial: Stem Cell Stories 1998–2008. *Science as Culture* 17 (1): 1–11.

Goodyear, Dana. 2016. The Stress Test: Rivalries, Intrigue, and Fraud in the World of Stem-cell Research. *New Yorker*, Feburary 29.

Gordon, Andrew. 2012. *Fabricating Consumers: The Sewing Machine in Modern Japan*. Berkeley, CA: University of California Press.

Gottweis, Herbert, Brian Salter and Catherine Waldby. 2009. *The Global Politics of hESC Science: Regenerative Medicine in Transition*. New York: Palgrave Macmillan.

————. 2010. Explaining Hwang-Gate: South Korean Identity Politics Between Bionationalism and Globalization. *Science, Technology & Human Values* 35 (4): 501–524.

Grose, Simon. 2008. Australian Agency Proceeds Cautiously with Hwang Patent. *Nature Medicine* 14 (12): 1300.

Hoffman, Michael. 2014. The Truth Is, We Have Gotten Too Used to Lying. *Japan Times*, March 29. http://www.japantimes.co.jp/news/2014/03/29/national/media-national/the-truth-is-we-have-gotten-too-used-to-lying/#.V00c1fkwiUk. Accessed 27 May 2016.

Hong, Sungook. 2008. The Hwang Scandal that 'Shook the World of Science'. *East Asian Science, Technology, and Society Journal* 2 (1): 1–7.

Hwang, Woo-suk, Choi Jaechon and Kim Byungjong. 2004. *Na ui Saengmyeong Iyagi*. Hyohyeong Chulpan.

Jong, Lee Tee. 2006. Why Does Disgraced Scientist Hwang Woo Suk Still Make the Hearts of Women Everywhere Go a-Flutter? *The Straits Times*, Singapore. Lifestyle Section, January 15.

Kayukawa, Junji. 2014. STAP Saibo Jiken ga Boukyaku Saseta koto. *Gendai Shisoo* 42 (12): 84–99.

Kim, Huiwon. 2006. Gwahangnyuseu moniteoring bangan mosaek jung, gija insikbyeonhwa neun? Geulsse. *Sinmum gwa bangsong*, November, 122–25.

Kim, Geun Bae. 2007. *Hwang Woo-suk Sinhwa wa Daehan Minguk Gwahak*. Yeoksa Bipyeong-sa.

Kim, Jongmyoung. 2007. 'Hwangppa' Hyeonsang ihaehagi: eum-mo ui mun-hwa, chaek-im jeonga cheongchi. *Hanguk Sahoehak* 41 (6): 76–111.

Kim, Tae-Ho. 2008. How Could a Scientist Become a National Celebrity?: Nationalism and the Hwang Woo-suk Scandal. *East Asian Science, Technology, and Society Journal* 2 (1): 27–45.

Kim, Yumi. 2008. Gungmin 88.4% Hwang Woo-suk Yeongu jaegae chanseong. *Joong-ang Ilbo*, July 20. http://news.joins.com/article/3231575. Accessed 27 May 2016.

Kim, Jongmyoung. 2009. Public Feeling for Science: The Hwang Affair and Hwang Supporters. *Public Understanding of Science* 18 (6): 670–686.

Kim, Gyeongnye. 2010. Hwang Woo-suk satae reul tonghaeseo bon saeng myeong uiryo gisul gwa jendeo. *Hyeondae-sahoe gwahak yeongu*, 169–88.

Kim, Leo. 2011. "Your Problem Is that Your Face Reveals Everything When You Are Lying": The Making and Remaking of Conduct in South Korean Life Sciences. *New Genetics and Society* 30 (3): 213–225.

Kim, Sang-Hyun. 2014. The Politics of hESC Research in South Korea: Contesting National Sociotechnical Imaginaries. *Science as Culture* 23 (3): 293–319.

Kim, Segyun, Gapsun Choi, Seongtae Hong, et al. 2006. *Hwang Woo-suk Sataewa Hanguk Sahaoe*. Nanam Chulpan.

Kimura, Ikuko. 2014. 'Rikejyo' Obokata san no 'shiroi kappogi', ninki kyuu jyooshyoo. *Sankei West*, February 5. http://www.sankei.com/west/news/140205/wst1402050064-n1.html. Accessed 26 May 2016.

Laidlaw, James. 2010. Agency and Responsibility: Perhaps You Can Have too Much of a Good Thing. In *Ordinary Ethics: Anthropology, Language, and Action*, ed. Michael Lambek, 143–164. New York: Fordham University Press.

Lancaster, Cheryl. 2016. The Acid Test for Biological Science: STAP Cells, Trust, and Replication. *Science and Engineering Ethics* 22: 147–167.

Lee, Sanghwa. 2005. *Hwang Woo-suk ui Kkum: gye-ran euro pawi rul chija! Haneul eul gamdong sigija!*. Deongseo munhwa-sa.

Lee, Hyeonjeong. 2016. Hwang Woo-suk 1-beon pae-a julgisepo gukga paejulgisepo lo deungnok. *Seoul Sinmun*, November 16, 11.

Leem, S. and J. Park. 2008. Rethinking Women and Their Bodies in the Age of Biotechnology: Feminist Commentaries on the Hwang Affair. *East Asian Science, Technology, and Society* 2 (1): 9–26.

Mandavilli, Apoorva. 2005. Profile: Woo-Suk Hwang. *Nature Medicine* 11 (5): 464.

Mikami, Koichi. 2015. State-Supported Science and Imaginary Lock-in: The Case of Regenerative Medicine in Japan. *Science as Culture* 24 (2): 183–204.

Misaki, Tomeko. 2012. Sengo Joseino Chakui Kappogi to Shiroi Apron: Bundansareru Shintai Renzokusuru Bosei. In *Chakuisuru Shintai to Joseino Syuenka*, ed. Sachiko Takeda. Kyoto: Shibunkaku.

Obokata, Haruko. 2016. *Anohi*. Tokyo: Kodansha.

Oh, Cheol-u. 2011. Hwang Woo-suk ui koyote bokje kwaegeo ga ju neun silmanggam. *Saieonsu On*, October 18. http://scienceon.hani.co.kr/31847. Accessed October 2016.

Park, Jaeyung, Hyoungjoon Jeon and Robert A. Logan. 2009. The Korean Press and Hwang's Fraud. *Public Understanding of Science* 18 (6): 653–669.

Parry, Jane. 2005. Korean Women Rush to Donate Eggs After Research Pioneer Resigns. *British Medical Journal* 331 (7528): 1291.

Salter, Brian and Charlotte Salter. 2007. Bioethics and the Global Moral Economy: The Cultural Politics of hESC Science. *Science, Technology, & Human Values* 32 (5): 554–581.

Sato, Aki. 2014. Obokata san kouka? Kappogi ga buumu chyuumon 3 bai no mise mo. *Asahi Shimbun Digital*, February 6. No Longer Available. Accessed January 2015.

Sato, Takahiko. 2015. *STAP Saibo Nokosareta Nazo*. Tokyo: Seiunsha.

Shin, Seung-cheol. 2005. *Sesang eul bakkun gwahakja Hwang Woo-suk.* Ja-eu gwa mo-eum chulpan-sa.

Simmel, Georg. 1969. *The Sociology of Georg Simmel.* Translated by Kurt Wolff. New York: The Free Press.

Sleeboom-Faulkner, Margaret. 2008. 'Debates on Human Embyronic Stem Cell Research in Japan: Minority Voices and Their Political Amplifiers. *Science as Culture* 17 (1): 85–97.

———. 2011. Regulating Cell Lives in Japan: Avoiding Scandal and Sticking to Nature. *New Genetics and Society* 30 (3): 227–240.

———. 2014. *Global Morality and Life Science Practices in Asia: Assemblages of Life.* New York: Palgrave Macmillan.

Sleeboom-Faulkner, Margaret and Seyoung Hwang. 2012. Governance of Stem Cell Research: Public Participation and Decision-Making in China, Japan, South Korea, and Taiwan. *Social Studies of Science* 42 (5): 684–708.

Staff. 2014a. Obotaka-shi Kaiken Sokuho—Kappogi wa san nen mae kara kite jikken shiteita, *The Page*, April 9. https://thepage.jp/detail/20140409-00000018-wordleaf. Accessed 26 May 2016.

———. 2014b. Obokata-shi Kaiken Raibu [8] Zero dewa naku mainasu 100 kara kenkyu ni mukiau. *Sankei News*, April 9. http://www.sankei.com/life/news/140409/lif1404090030-n1.html. Accessed 26 May 2016.

———. 2016. Obokata Questioned Over Alleged Theft of Riken Stem Cell Samples. *Japan Times*, February 18. http://www.japantimes.co.jp/news/2016/02/18/national/crime-legal/obokata-questioned-alleged-theft-riken-stem-cell-samples/#.V3zLvesrKUk. Accessed 20 June 2016.

Tanimoto, Mayumi. 2014. Hitobanjyuu nakiakashita 30sai wakate jyosei kenkyushya to kaku wagakuni niwa goshippu shinbun shika nai rashii. *WirelessWire News*, January 31. https://wirelesswire.jp/london_wave/201401310211.html. Accessed 26 May 2016.

Thompson, Charis. 2013. *Good Science: The Ethical Choreography of Stem Cell Research.* Cambridge, MA: The MIT Press.

Tsuge, Azumi and Hyunsoo Hong. 2011. Reconsidering Ethical Issues About "Voluntary Egg Donors" in Hwang's Case in Global Context. *New Genetics and Society* 30 (3): 241–252.

Wirten, Eva Hemmungs. 2015. The Pasteurization of Marie Curie: A (Meta) Biographical Experiment. *Social Studies of Science* 45 (4): 597–610.

Youn, Sugmin, Namjun Kang and Heejin Kim. 2009. The Never-Ending Myth: An Analysis of the Sociopsychological Mechanism of Hwang Woo-Suk Syndrome. *Korea Journal* x: 137–164.

Marcie Middlebrooks is a medical anthropologist interested in bioethics, science, and religion in East Asia. She has been a Fulbright Scholar in Japan and received a master's degree in Religious Studies from Seoul National University. Her Ph.D. research at Cornell University focused on Buddhist responses to the Hwang Woo-suk stem cell research scandal in South Korea.

Hazuki Shimono is a Japan Society for the Promotion of Science Research Fellow and a Ph.D. candidate in the School of Humanities and Sociology at the University of Tokyo. She has a degree in the Theory and History of Architecture from Columbia University (Barnard College) and has written articles in leading journals including the *Journal of History of Science, Japan*.

Part II

Therapeutic Horizons

Establishment and Use of Injectable Human Embryonic Stem Cells for Clinical Application

Geeta Shroff

Spinal cord injury (SCI) is a devastating and challenging neurological ailment that affects millions across the world[1]. Generally caused by sports, car accidents, tumors, falls, or infection, SCI leads to paralysis and loss of sensory and motor functions and sometimes is accompanied by urinary, cardiac, and respiratory dysfunctions. SCI is associated with permanent disability and decreased life expectancy. The poor regenerative capacity of the adult spinal cord results in severe sensory and motor deficits. Presently, there is no cure. It impacts the patient physically, psychologically, and socially and financially.

A systemic review estimated the incidence rate of traumatic SCI in Asia as ranging from 12.06 to 61.6 per million in people between the ages of 26 and 56. A recent review of SCI epidemiology in developing countries reported the incidence to be 25.5 million cases per year (Rahimi-Movaghar et al. 2013). In India alone, approximately 1.5 million

G. Shroff (✉)
Nutech Mediworld, New Delhi, India

© The Author(s) 2018
A. Bharadwaj (ed.), *Global Perspectives on Stem Cell Technologies*,
https://doi.org/10.1007/978-3-319-63787-7_5

live with SCI, and 10,000 new cases are added each year (Gupta et al. 2008). The neurological recovery in patients with traumatic SCI, as evaluated by the American Spinal Injury Association (ASIA) is low (6–13%), and only about 2.1% of these patients have been reported as gaining any functional strength (Kirshblum et al. 2004). According to ASIA, the severity of an injury is categorized as either complete or incomplete: a complete injury is the complete absence of sensory and motor function below the level of injury. If there are some preserved motor and sensory functions below the level of injury, the case is diagnosed as incomplete SCI (Grossman et al. 2012; Dalbayrak et al. 2015).

Human embryonic stem cells (hESCs) are self-renewing cells with a potential to differentiate into all types of human cells. hESCs are able to replicate indefinitely, differentiate into all three primary germ layers cell lines, and are karyotypically stable (Erceg et al. 2008; Keirstead et al. 2005; Ware et al. 2014; Shroff 2005).

These cells have the potential for cell replacement and regeneration therapies for human diseases. hESCs were derived and characterized as early as 1982 from fresh or frozen cleavage-stage donated human embryos produced by in vitro fertilization (IVF) (Shroff et al. 2014). The viable cell lines were obtained from the inner cell mass or blastocyst. hESCs have also been derived and established from single blastomeres of the four- or eight-celled embryo and 16-celled morula. Since then, a plethora of research has indicated that hESCs can be used for various diseases like diabetes; liver, autoimmune, and immune disorders; Parkinson's disease; Alzheimer's disease; age-related macular degeneration; and SCI.

Despite its great potential in treating clinical conditions such as SCI, hESCs have not been used extensively in humans. This is largely due to the technology which advocates hESC lines and a lack of knowledge about their use. Furthermore, hESC cell lines have shown chromosomal and genomic instabilities, with acquisition of loss of heterozygosity or copy-number variation in cancer-related genes. hESCs have also been associated with teratoma formation and fear of being immunologically rejected. There have been safety concerns and challenges in the use of hESCs. The use of animal feeder cells leads to cross contamination. There is a risk of xenogeneic pathogen cross-transfer and other unknown substances capable of eliciting a detrimental immune response in transplanted hosts. There are ethical issues which center around the repeated need for blastocyst to create the cell lines.

In this chapter, I will describe the establishment and transplantation of injectable hESCs for chronic SCI. I also will illustrate the working of hESC through case-study data and discuss how this unique therapeutic platform has achieved significant success in treating chronic SCI.

Literature

Several studies in the past two decades have researched cell-based therapies for SCI. The replacement of damaged neural tissues and reestablishing connections between the central and peripheral nervous systems are vital for SCI treatment strategies, and cells with a potential of self-renewal and the ability to differentiate into multiple cell types would be best suited for SCI patients. Park et al. (2005) performed autologous bone marrow cell transplantation at the injury site in conjunction with the administration of granulocyte-macrophage colony-stimulating factor in five patients with complete SCI and followed up for 6–18 months. Overall, three patients improved from ASIA scale A to C, one improved from ASIA scale A to B, and one did not show any improvement. None of the patients, however, showed any serious complications. Lima et al. (2010) transplanted olfactory mucosa autografts in seven patients ranging from 18 to 32 years of age and with an ASIA scale of A. Every patient had improved ASIA scales, and two of the patients had moved to ASIA scale C by the end of treatment. In another study, a 37-year-old female SCI patient was transplanted with HLA-matched human cord blood cells at the site of injury. Investigators observed improved sensory perception and movement in the hips and thighs 41 days after the transplantation. Regeneration of the spinal cord at the injured site was observed in the computed tomography and MRI scan. Kang et al. (2005) recommended that hESCs' transplantation protocols should encourage the use of human material, as animal components carry a risk of xenogeneic pathogen cross transfer.

A Phase I hESC human clinical trial was approved by the FDA in 2009. Popularly referred to as the Geron trial, it was abandoned midway due to financial constraints (Lukovic et al. 2014). The cells in the Geron trial also contained animal components such as the B27 supplement or Matrigel. Asterias Biotherapeutics have bought the rights to Geron to conduct a human clinical trial, approved by the FDA (Leuty 2014). Ocata is also developing hESC-based therapies for various disorders; initial results in patients with macular degeneration have been promising. A number of studies have been conducted in animal models to observe the capabilities of stem cells in improving the motor functions in SCI patients. In transplanted placenta-derived mesenchymal stem cells to rat models with SCI, these rats showed significant improvements in their motor and hind-limb functions after three weeks of study. The Basso, Beattie, and Bresnahan scale also shifted from 2 to 13 within three weeks of treatment (Sharp et al. 2010). Kerr et al. (2003) studied the human pluripotent stem cell derivatives transplanted to rat models to observe the improvement in the motor functions and observed significant improvements in hind-limb locomotion as compared with controlled animals in 12 weeks of study.

Sharp et al. used hESC-derived oligodendrocyte progenitors in adult cervical contusion rat models and observed that transplanted hESC-derived OPCs survived even after the nine-week study period. Rossi et al. (2010) observed that transplanted animals had an improved functional outcome with an early recovery rate of balance and coordination and skilled forelimb movement when human motor neuron progenitor cells derived from hESCs were transplanted in rats with SCI. In addition to the evaluation of stem cells in animal models, early-phase clinical trials regarding the efficacy of stem cells in SCI yielded mixed results. Yoon et al. (2007) conducted a non-randomized Phase I/II clinical trial to treat SCI with transplantation of bone marrow cells and observed a significant change in the ASIA scores of patients in the acute and sub-acute treatment groups although no improvement was observed in the chronic treatment group. In Mackay-Sim et al.'s (2008) three-year clinical trial of 12 paraplegic patients, olfactory ensheathing cells (OECs) were transplanted through multiple routes and no changes were made to ASIA scores, and no patients experienced improvements to neurological and functional levels. Lima et al. performed a pilot scale clinical study in seven patients with chronic SCI. They transplanted the OEC to the SCI patients through surgical mode. The authors found significant changes in the MRI observations and ASIA score. The bowel and bladder movement of patients were also improved.

We have previously reported improvements in some of our SCI patients with SCI after undergoing hESC therapy (Shroff and Gupta 2015). In another study, we have reported improvement in bowel and bladder sensation and control (Shroff 2015c). Patients with acute SCI were not included, to rule out the natural recovery of disease. All the patients in our study had SCI for more than one year and had not benefitted from any other treatment. Bretzner and his colleagues (2011) state that chronic SCI patients are more acceptable for hESC-derived transplantations as compared to acute patients, as the former are 'less likely to suffer opportunity costs from study participation', an important ethical consideration when 'knowledge value', not 'therapeutic benefit', motivates research.

The standard treatment for chronic SCI includes high doses of steroids (methylprednisolone) and immunosuppressants. The mild therapeutic effect of methylprednisolone is associated with a number of other side effects (Willerth and Sakiyama-Elbert 2008; Bracken et al. 1990; Bracken 2012). Hugenholtz states there are no evidence-based standards regarding the use of high doses of methylprednisolone for SCI treatment (Kirshblum et al. 2011), and surgery for SCI patients does not show any improvement between treated and non-treated patients (Ronaghi et al. 2010). Thus, very few or negligible treatment options are available.

The Breakthrough

Since 2000, I have researched and developed a unique in-house patented technology to culture and maintain hESCs in our GMP-, GLP-, and GTP-compliant laboratory. In particular, I have studied very small stem cells (VSELSCs) of pre-blastomeric origin derived from a two-celled stage fertilized egg. These cells (0.7–1.5 μm) known as blastomeric-like express pluripotency genes and differentiate into cell types from all the germ layers.

These cells were taken from a fertilized ovum discarded during a regular IVF cycle, with full donor consent. The cell-culture technique produces an hESC line free from animal products, feeder layers, growth factors, leukemia inhibitory factor, supplementary mineral combinations, amino acid supplements, fibroblast growth factor, membrane-associated steel factor, soluble steel factor, and conditioned media (US Granted Patent No US 8592, 208, 52). The hESC line was characterized for its pluripotent nature, its differentiation into neuronal lineages, and its use for the treatment of neurological disorders. We have characterized the hESC at the molecular, cellular, and functional levels using scanning electron microscopy, transmission electron microscopy, confocal microscopy, reverse transcriptase polymerase chain reaction, and flow-cytometry analysis. The cell line has also been characterized based on its long-term proliferation and maintenance, karyotyping, and in vivo differentiation as teratoma formation assay. It has been chromosomally stable since the year 2000 and for >4000 passages. The study also provides a composition of injectable stem cells in a ready-to-inject form that is simple to prepare and safe, cost effective, efficient, easily transportable, scalable, and with a shelf life of greater than 6 months.

The safety and efficacy of the cell line has been established (Shroff et al. 2015b). We have used these cells to treat over 1500 patients with diverse ailments including diabetes, myocardial infarction, cerebral palsy, SCI, Lyme disease, spinocerebellar ataxia, Friedrich's ataxia, Alzheimer's disease, Parkinson's disease, autism, and cerebral palsy (Shroff et al. 2014, 2015, 2015d; Shroff 2015). No teratoma formation has been observed. The hESCs were obtained from a one-time harvest at the pre-blastomeric stage. The cell line thus developed was created from a single expendable fertilized ovum 24 hours after fertilization. With no animal products in the media. We have developed a simplified cell-culture system free of exogenous cells and supplements of animal origin for expansion of hESCs in a substantially undifferentiated state.

The study was approved by an independent ethics committee. The study was conducted in accordance with the Declaration of Helsinki in a GCP-compliant condition. Each patient provided verbal, written, and video consents; the ethics committee approved this process. In addition, the cells are cultured and maintained per our in-house patented technology (United States Granted Patent No US 8592, 208, 52) in our laboratory certified as being compliant with good manufacturing, good laboratory, and good tissue practices.

Patient data was validated by Moody's International (document number NH-heSC-10–1), GVK Biosciences (NM-Hesc-10–1, 18 November 2010), and Quality Austria Central Asia (document number QACA/OCT/2013/26). These companies were allowed to examine the medical and statistical data at the institute and were able to meet the patients.

Case Studies

Here, I present our data on paraplegic and quadriplegic patients treated with hESCs. All the patients were scored on the basis of a scale developed by ASIA before and after treatment (ASIA/IMSOP, 1996). After the treatment, all three patients showed significant improvement in their sitting balance, bowel and bladder control and sensation, and power and movement of lower and upper limbs. No adverse events were reported. The treatment strategy was divided into four phases. In the first, T1 (eight weeks for paraplegics, 12 for quadriplegics), 0.25 mL (<4 million cells) hESCs were administered intramuscularly twice daily to 'prime' the body and to prevent the recipient's immune system from rejecting the stem cells. Every 10 days, 1 mL hESCs (<16 million cells) were administered intravenously to 'home in on' the required area. Every 7 days, 1–5 ml were administered via any of supplemental routes (brachial plexus block, intrathecal, caudal, epidural, popliteal block, and/or deep spinal muscle and epidural catheter) to introduce the stem cells as closely to the injured site as possible (local action). After a gap period of 4–8 months, successive phases like T2 (four to six weeks) and T3 (four to six weeks) used the T1 dosage regime; treatment was repeated annually if needed. This protocol was developed on the basis of a pilot study conducted on 72 patients that found the extension of a treatment period more than eight weeks in paraplegics and more than 12 in quadriplegics do not lead to better results. A gap of 4 months between the subsequent treatment phases was determined to allow the injected hESCs to develop into mature cells and regenerate the affected parts. T2 and T3 treatment periods were incorporated to add more cells into the body, thus allowing more repair and regeneration. Biochemical and radiological investigations were completed before the start of the treatment and at regular intervals. In-house physicians and nurses carefully observed patients for antigenic or ana-phylatic responses.

Patient 1

An Australian patient with quadriplegia for 14 years was admitted to Nutech Mediworld on 8 March 2008. Patient history revealed he had suffered a major trauma to the neck while playing rugby, which resulted in injury to the spine at the C1 and later receded to the C2 level. His investigator assigned an ASIA score of an A.

At the time of admission, the patient was unable to move his upper and lower limbs and suffered from a complete loss of sensation except on his face. He was on ventilator support with tracheostomy at 17 breaths per minute, and speech was co-incident with the ventilator. He had no sitting balance, and the plantar and abdominal reflexes were absent with an exaggerated ankle jerk. His lower limb had clonus, and he had no deep sensation. In addition to having no bladder or bowel control, he needed three full-time caretakers and could not eat more than one meal a day. Magnetic resonance imaging tractography showed the visualization of nerve fibers/tracts in the upper cervical cord from the cervicomedullary junction caudally up to the C2; cord fibers were not discerned up to the D1 (Fig. 1).

The patient underwent four sessions of hESC therapy. After his treatments, the patient was weaned off his ventilator and was able to remain from it for up to 12 hours. He was able to freely move his neck, shrug his shoulders, and show movement of his arms and hands. His sitting balance improved significantly, and he could stand with a chest orthosis and Hip Knee Ankle Foot Orthosis. His deep sensation was increased up to the abdomen. His post-treatment ASIA score was a C, and his last follow-up was 8 November 2013 (Fig. 2).

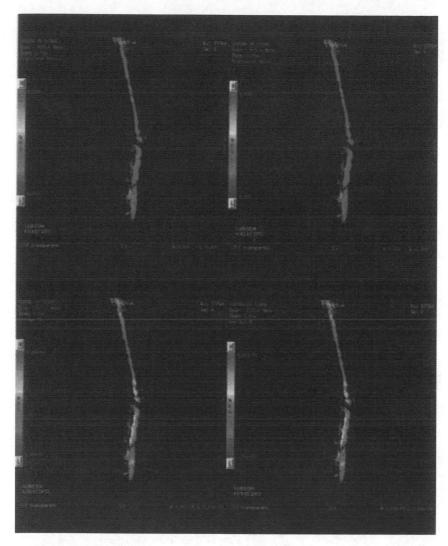

Fig. 1 Trachtography

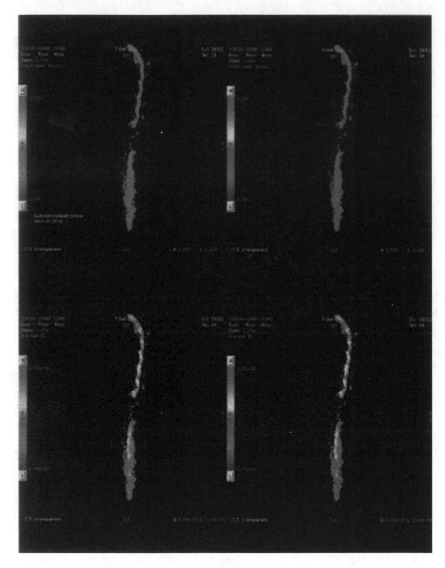

Fig. 2 Patient 1 after treatment

Patient 2

A 28-year-old Indian male was admitted to Nutech on 8 October 2011, with complaints of loss of bowel and bladder control, a left lower limb paralyzed with intact sensation, and right lower-limb movements with no sensation. The patient had had an accident in October 2010, sustaining D11 and D12 fractures and a right ulnar fracture. He underwent surgical procedure with instrumentation at the D9-L1 level and after surgery had regular sessions of physiotherapy and occupational therapy, which helped improve motor functions in the right lower limb and sensation in the left lower limb. He was able to lift and hold his right lower limb and could hop on crutches but was unable to walk. His MRI showed a fracture of the D11 vertebral body with anterior subluxation of D10 over D11, compression fracture of D12 vertebral body, decreased spinal size at D11 due to fracture, subluxation with indentation of the cord at D10-D11, and minimal extradural hematoma at D10–D11. Cord contusion was noted at the D10–D11 level, and a D11 and laminar fracture was seen on the right side. His ASIA score was an A (Fig. 3).

The patient underwent two sessions of hESC therapy. After the first treatment, he showed improvement in bladder and bowel sensation with partial control, improvement in sensation in his left and right lower limbs (he could bend his knees and hold that position). He showed a decrease in clonus and jerks in the lower limb and in calf pain. He was able to walk with Ankle Foot Orthosis calipers and walkers and could climb stairs with little help.

After a gap of more than two years, the patient had another treatment. During therapy, he was also given supplements, including calcium (Calcitriol 500 mg × OD), 15 units of B-complex, ferrous fumarate and folic acid OD, cholecalciferol sachets (one per week), and nectra powder. This treatment protocol resulted in improved lower-limb strength, improved left lower-limb sensation at the foot-planar and dorsal aspects, and improved calf and gait with a quadripod stick and bars. The right thigh and toe movements were stronger than before. His bowel and bladder control improved with full voiding and evacuation sensation. The patient's last follow-up was 30 October 2013.

(b) After Treatment

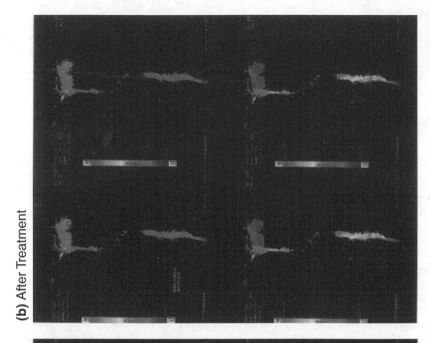

(a) Before Treatment

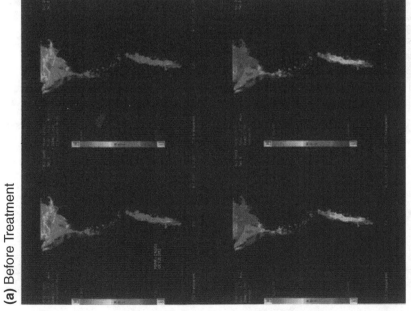

Fig. 3 Tractography images of Patient 2 before and after treatment

Patient 3

A 36-year-old female from the USA was admitted on 1 February 2010. She had been healthy until 2007, when she had an accident, sustaining injury at C6-C7. She underwent surgical fixation of C5-T1 with rods. She had been wheelchair bound on a quad support since that time, and her MRI revealed that a spiral fracture of the right humerus was surgically fixed with titanium plates and screws. She also had no movement of the B/L lower limb, slight movement in the B/L upper limb, absence of sensation below the sternum, no bowel or bladder sensation or control, pain from waist to toe, neuralgic pain below her waist, the inability to grasp any object, poor sitting balance, inability to walk, and dizziness on sitting up. On examination, she had claw hand, her thumb movement was absent from the right hand and weak in the left, and she had no hand function, a weak grip, and had poor pelvic control in the quadruped position. Her ASIA score was an A.

The patient was treated with hESC along with extensive physiotherapy and occupational therapy. She underwent six sessions of hESC therapy with a gap phase between sessions varying from 4–11 months, with her last follow-up on 23 December 2013. Following the treatment, she showed improvement in upper-limb muscle strength, contraction of B/L hips and knees, and an improvement in posture, with good sitting balance. The patient had developed a good standing balance with calipers, good grasping with her right hand, good thumb movement in both hands, and a good pinch and release. She was able to write, make a near-normal fist, had developed sensation until B/L knees and at the sole of the foot, and had appreciable extension in both knees; she also could stand for long durations and even take steps with the Knee Ankle Foot caliper and binder. She also showed improvement in bladder sensation (could hold the speed of voiding) and bowel sensation (pushed while passing motions). She was able to crawl independently and stand from a sitting position while holding onto bars. Her post-treatment ASIA score was a B.

We can assume the hESCs in our study may have followed the pattern of homing in at the injured sites and regenerating the affected regions, as discussed previously. We cannot rule out the possibility that in our study, the hESCs may have differentiated into neurons and helped in tissue repair at the site of injury in the spinal cord and regenerated the cells for improvement in the functioning of the spinal cord.

Dramatic improvements in the health of all three patients we studied were observed following the treatment with hESC in our study. None of the patients developed teratomas and were not given any immunosuppressants. hESCs could be an effective and safe therapeutic option for treatment of patients with SCI. However, there is a need to conduct clinical trials on large number of patients with SCI to determine the safety and efficacy of this treatment.

A Retrospective Study

Nutech Mediworld has evaluated the efficacy and safety of hESC therapy in 226 patients with SCI (Shroff 2016). In the first treatment phase (T1), 0.25 mL hESCs were administered intramuscularly twice daily (1 mL every 10 days intravenously and 1–5 mL every 7 days). Of the 153 patients on the ASIA scale A at the beginning of T1, a significant number of patients ($n = 80$; 52.3%) had moved to lower scales at the end of T1 ($p = 0.01$). At the end of T2, of the 32 patients on ASIA scale A, 12 patients (37.5%) had moved to scale B ($p = 0.01$). Of 19 patients, three (37.5%) had moved to scale B at the end of T3 ($p = 0.02$). No serious adverse events were observed.

Study Design

The data of a single cohort of SCI patients treated with hESCs conducted from 24 May 2005 to 31 August 2012, at a single site in New Delhi, India, and were collected retrospectively. In the initial two years (2002–2004), the safety of hESC therapy was assessed in 33 patients (not included in this analysis) with various incurable diseases. Thereafter, efficacy of the therapy, dose schedule, and protocol for administration of hESCs and therapy schedules were established in a pilot study conducted on 72 patients. After that, a study (validated by GVK Biosciences) was conducted on 108 SCI patients verifying the safety and efficacy of hESC in SCI patients (not included in the present analysis). The present study of 226 SCI patients was undertaken after these two studies. The same protocol was followed in the group of patients analyzed in this study.

The study was performed under the proper supervision of a team of physicians that included external consultants and was validated by an external clinical research organization. The patients were scored per ASIA scale30 by independent physicians before and after the treatment and by in-house physicians and the rehabilitation team. After confirmation of diagnosis, patients were tested for hypersensitivity reactions with hESCs (0.05 mL injected subcutaneously). The three treatment phases were separated by gap periods so the hESCs could grow, repair, and regenerate the affected parts.

Patients with a documented diagnosis of SCI elsewhere of fewer than 3 months before the start of therapy were included in this study. All these patients had undergone other treatment(s) such as physiotherapy, occupational therapy, and so forth before coming to our center. Patients with acute SCI were excluded, to rule out results of the body's natural healing abilities. Patients who were pregnant, lactating, or confirmed to have received other forms of cell therapies within the last 12 months of treatment were not accepted. All patients provided written and video informed consents before the treatment began.

Patients then entered the first treatment phase (T1): eight weeks for paraplegics and 12 for quadriplegics, wherein 0.25 mL (<4 million cells) hESCs were administered intramuscularly twice daily to 'prime' the body and prevent the recipient immune system from rejecting the stem cells. In addition, every 10 days 1 mL hESCs (<16 million cells) were administered intravenously to home in on the required area and for systemic reach. To introduce the stem cells as closely to the injured site as possible (local action), 1–5 mL hESCs (depending on the route of administration) were administered every 5 to 7 days by any of the supplemental routes (Fig. 4).

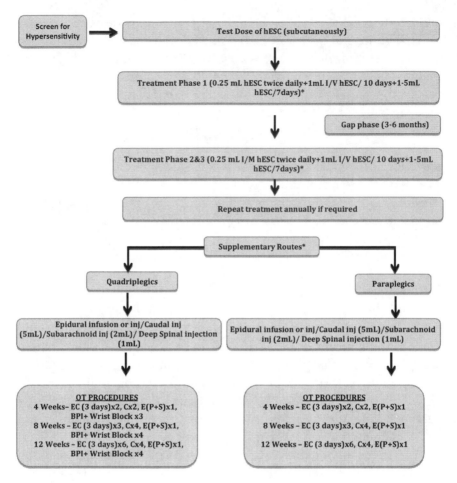

Fig. 4 Treatment plan for Spinal Cord Injury

The duration of treatment and gap phases varied in quadriplegic patients and paraplegic patients, as quadriplegics are generally more difficult to treat (Paralysis, Paraplegia, and Quadriplegia. MD guidelines 2015). After a gap period of 4–8 months, patients entered the subsequent treatment phases (T2 and T3), in which they were administered the same dosage regime as in T1. Each treatment phase lasted four to six weeks and was 4–8 months apart. In T2 and T3, an additional dose of hESCs was administered through any of the supplemental routes.

No immunosuppressants were given to the patients, who also received physiotherapy and/or occupational therapy. Rehabilitation focused on patients' overall improvement, and patient mobility was encouraged by using different ambulatory aids depending on the requirement (e.g., a patient with paraplegia was made to stand with full support on a hip-knee-ankle-foot orthosis, and as connectivity was regained, the support was reduced to a knee-ankle-foot orthosis, then knee brace and ankle support, and then ankle support). The walking aids were gradually reduced from manual support and walker to walker to crutches to walking stick to, finally, no aid.

Assessment

Each patient's pre-therapy status was assessed at admission. The percentage with changes or no changes were calculated after each session of the therapy and reported. Statistical tests or tests of significance were performed.

Data Validation

Patient data was validated by Moody's International (document number NH-hESC-10-1), GVK Biosciences (NM-Hesc-10-1, 18 November 2010), and Quality Austria Central Asia (document number QACA/OCT/2013/26). These companies examined the medical and statistical data present at the institute and met the patients.

Statistical Analysis

Descriptive statistics were performed to summarize data. SPSS software version 19.0 (IBM Corporation, Armonk, NY) was used for the data analysis. A chi-squared test was used to compare the AIS score at baseline and at the end of the therapy. A p value of < 0.05 was considered significant (Table 1).

Table 1 Change in American Spinal Injury Association scales of patients (overall) from admission to discharge at the end of each treatment period

Baseline characteristics	End of the treatment					p value
	ASIA scale					
	A; no. (%)	B; no. (%)	C; no. (%)	D; no. (%)	E; no. (%)	
T1 (n = 226)						0.02
A (n = 153)	73 (47.7)	23 (15)	57 (37.3)	–	–	
B (n = 32)	–	18 (56.3)	13 (40.6)	1 (3.1)	–	
C (n = 36)	–	–	31 (86.1)	5 (13.9)	–	
D (n = 5)	–	–	–	2 (40.0)	3 (60.0)	
T2 (n = 58)						0.01
A (n = 32)	20 (62.5)	12 (37.5)	–	–	–	
B (n = 9)	–	7 (77.8)	2 (22.2)	–	–	
C (n = 17)	–	–	17 (100)	–	–	
T3 (n = 19)						0.02
A (n = 8)	5 (62.5)	3 (37.5)	–	–	–	
B (n = 4)	–	4 (100)	–	–	–	
C (n = 7)	–	–	6 (85.7)	1 (14.3)	–	

ASIA American Spinal Injury Association

Results

A total of 226 SCI patients (paraplegic = 136, quadriplegic = 90) were included in the study. Overall, 203 patients had SCI due to trauma and 23 to miscellaneous causes like transverse myelitis (*n* = 4), Potts spine (*n* = 7), tumors (*n* = 3), and contusion (*n* = 9). The majority of these patients were men (167, 73.9%), and the mean age was 28 (range 20–34 years). Among paraplegic patients, 124 had complete injury, whereas among the quadriplegic patients, 71 had complete injury. The average days of treatment in T1 was 73 days for quadriplegic patients and 62 days for paraplegic patients, and the average gap period was 122 days for quadriplegic patients and 136 days for paraplegic patients.

All patients started intensive dosing, and 50 were present in all study periods. Overall, patients who discontinued the study for various reasons cited personal status and financial reasons (39%), satisfaction with their progress and cure (32%), dissatisfaction with their progress (5%), and returning for treatment after a long gap (24%, i.e., after 31 August 2012, these patients were not part of this analysis).

Efficacy Evaluation

Changes in the ASIA impairment scale from admission to discharge at the end of each treatment period are presented in Table 1. Of the 153 patients in ASIA scale A at the beginning of T1, a significantly higher number of patients (n = 80, 52.3%) had moved to lower scales at the end of T1 (p = 0.02). At the beginning of T2, 32 patients were in ASIA scale A; of these, 20 patients remained in scale A and 12 (37.5%) had moved to lower scales by the end of T2 (p = 0.01). Of the 19 patients at the start of T3, eight patients were in ASIA scale A. At the end of T3, three of these patients (37.5%) had moved to scale B (p = 0.02). The improvement in scales at the end of T1 and T2 is shown in Fig. 5. At the end of T1, 45% of the patients had improved by at least one ASIA grade. At the end of T2, 58% of the patients had improved by at least one grade, and at the end of T3, 70% of the patients had improved by at least one grade (Table 2).

MRI scans for 65 of patients and tractographies for 25 patients were conducted before and after therapy. Improvements were observed in the MRI tractography images of these patients before and after the therapy (Fig. 1 and 2).

Table 2 Change from baseline to last period in total American Spinal Injury Association scores by extent and level of injury

Study period	Results	ASIA grades No. of patients (%)
End of T1 (n = 226)	Improved by 1 ASIA scale	102 (45)
	Stationary	124 (55)
	Not improved	–
End of T2 (n = 58)	Improved by 1 ASIA scale	62 (58)
	Stationary	44 (42)
	Not improved	–
End of T3 (n = 19)	Improved by 1 ASIA scale	35 (70)
	Stationary	15 (30)
	Not improved	–

ASIA American Spinal Injury Association

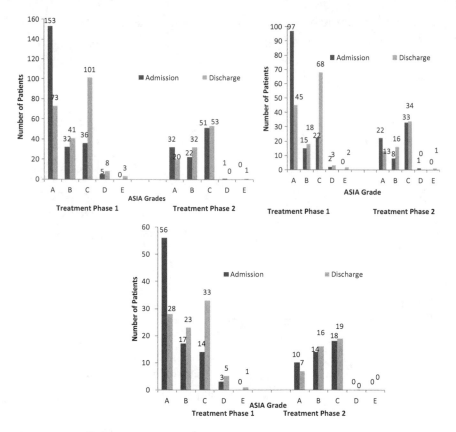

Fig. 5 Overall change in American Spinal Injury Association (ASIA) scale at the end of treatment phase 1 and 2

Paraplegics and Quadriplegics

Among the 136 patients with paraplegia, 97 were on ASIA scale A at the beginning of T1. Of these, a significant number of patients (52, 53.6%) had moved to lower scales by the end of T1 ($p < 0.05$). Of the 64 with paraplegia at the beginning of T2, 22 were on scale A. At the end of T2, nine patients (40.9%) had moved to scale B.

Among quadriplegic patients, 28 (50%) had shifted from scale A to the lower scales at the end of T1 ($p > 0.05$). Of the ten patients in scale A at the beginning of T2, three (30%) had improved and moved to lower scales. Of the three patients on ASIA scale A before T3, one (33.3%) had moved to scale B at the end of T3.

Gender Analyses

Of the 59 women in the study, 32 were on ASIA scale A at the beginning of the study. At the end of T1, 13 were on ASIA scale A and the rest had moved to lower ones. Among the 167 men, 121 were on ASIA scale A at baseline. At the end of T1, almost half ($n = 60$) remained on ASIA scale A, and another half had moved to lower scales. The improvement in scales at the end of T1, T2, and T3 is shown in Table 3. Gender was not a significant factor in the state of efficacy.

Table 3 Change in American Spinal Injury Association scales of patients (gender wise) from admission to discharge at the end of each treatment period

| | | | No. of patients | | | | |
| | | | At the end of treatment phase | | | | |
Gender	Treatment phase	Admission	A; no. (%)	B; no. (%)	C; no. (%)	D; no. (%)	E; no. (%)
Women	T1 ($n = 59$)	A ($n = 32$)	13 (41)	7 (22)	12 (38)	–	–
		B ($n = 10$)	–	7 (70)	2 (20)	1 (10)	–
		C ($n = 15$)	–	–	14 (93)	1 (7)	–
		D ($n = 2$)	–	–	–	1 (50)	1 (50)
	T2 ($n = 23$)	A ($n = 4$)	2 (50)	2 (50)	–	–	–
		B ($n = 6$)	–	4 (67)	2 (33)	–	–
		C ($n = 12$)	–	–	12 (100)	–	–
		D ($n = 1$)	–	–	–	–	1 (100)
	T3 ($n = 11$)	A ($n = 1$)	1 (100)	–	–	–	–
		B ($n = 4$)	–	3 (75)	1 (25)	–	–
		C ($n = 6$)	–	–	5 (83)	1 (17)	–
Men	T1 ($n = 167$)	A ($n = 121$)	60 (50)	16 (13)	45 (37)	–	–
		B ($n = 22$)	–	11 (50)	11 (50)	–	–
		C ($n = 21$)	–	–	17 (81)	4 (19)	–
		D ($n = 3$)	–	–	–	1 (33)	2 (67)
	T2 ($n = 83$)	A ($n = 28$)	18 (64)	10 (36)	–	–	–
		B ($n = 16$)	–	16 (100)	–	–	–
		C ($n = 39$)	–	–	39 (100)	–	–
	T3 ($n = 39$)	A ($n = 7$)	4 (57)	3 (43)	–	–	–
		B ($n = 12$)	–	12 (100)	–	–	–
		C ($n = 20$)	–	–	20 (100)	–	–

Table 4 Adverse events observed during each treatment period (safety population)

AE parameter; no. total AEs; no. (%)	T1 (n = 226) 57 (25.2)	T2 (n = 106) 9 (8.5)	T3 (n = 50) 6 (12)
Fever	23	3	1
Headache	15	5	3
Loose motions	3	–	–
Abdominal pain	2	–	1
Constipation	2	–	–
Itching	2	–	–
Pain	2	–	–
Fever and headache	1	–	–
Fever, anorexia, and hematuria	1	–	–
Headache and vomiting	1	–	1
Headache and vertigo	1	–	–
Weight loss	1	–	–
Nausea	1	1	–
Redness and itching	1	–	–
Acidity	1	–	–

AE adverse event

Safety Evaluation

No death or serious adverse events occurred during the study period. No teratoma formation was observed during or after the study. Adverse events observed during each treatment period are tabulated (Table 4). Mild fever was the most frequent adverse event during the study and resolved without sequel.

The present study is the first of its kind to demonstrate the adequate efficacy of hESC in SCI patients with a good tolerability profile. The results of the present study have given a new ray of hope to SCI patients. Given the improvement shown by our patients, we propose that hESC transplantation into SCI patients presents a unique opportunity to address this greatly unmet medical need.

Anesthesia and hESC Injectable Procedures

A major concern for physicians is hESC implantation with reduced pain. Nutech Mediworld has been using specialized procedures to implant hESCs in an SCI patient. Nutech Mediworld has thus developed a novel approach: the use of epidural and caudal routes.

An epidural route injects hESCs into the region outside the dura mater of the meninges, whereas hESCs implantation through the sacral membrane—approximately three centimeters above the tip of the coccyx and in continuum with the epidural space—is achieved by caudal route. The anesthesiologist plays a significant role in hESC transplantation by evaluating a patient's condition and protocol development along with suitable health-care management. This specialist encourages the optimal use of multimodal regimens as well as the implementation of novel techniques, ensuring improvement in pain control and the minimization of adverse events.

hESC implantation is not an easy task, as determination of the anatomic location of the implantation site is a major concern. High cell concentration within the region of interest is the physician's main target. However, we wish to prevent the dwelling of cells into other undesirable sites/organs. Intracoronary, transendocardial, transpericardial, intraventricular, intravenous, intramyocardial, and other routes of catheterization/administration have been previously reported. Further developments of catheterization systems for clinical studies involve other routes for administering hESCs such as intramuscular, intravenous, intrathecal, epidural, caudal, brachial plexus, popliteal, and/or deep spinal muscle injection. At our hospital, the epidural as well as caudal routes—procedures requiring anesthesiologists—have been used regularly for hESC implantation in SCI patients. The present study explains the caudal and epidural method of implantation.

Non-neuronal and neuronal cell lines—obtained from a single, spare, expendable, pre-implantation stage fertilized ovum taken during the IVF process, with due consent—are cultured, maintained, and stored in syringes for further use. The pre-filled, frozen syringes are thawed when required. The cells undergo quality analysis for determination of integrity, viability, and microbial contamination (Gupta and Barthakur 2014).

The paraplegic patients receiving caudal, epidural, and both procedures were the ages of 66, 20, and 43, respectively. Similarly, the quadriplegics receiving caudal, epidural, and both were 27, 12, and 23, respectively. The patients were under the care of skilled and experienced anesthesiologists and physicians. SCI patients being readied for hESC therapy were provided with the facts, implications, and consequences of the therapy. Each patient gave video and signed consents before treatment. In addition to understanding the general protocol and time involved in the procedure, patients were advised to discuss with their clinician and physician whether medications could be taken on the day of the injection.

The physician took a record of each patient's allergic reactions and medications, and patients were asked to avoid drinking and eating after midnight. Medicines were taken with a sip of water. Alcohol and smoking were restricted before, during, and after the procedure. Patients were asked to change into a hospital gown, to allow physicians to clean the injection area and easily visualize the injection site. After moving the patients to the operation theatre, the hESCs were transplanted with a procedure involving trained anesthesiologists.

Epidural Procedure

An 'epidural' injection or catheter infusion involves the region outside the dura mater of the meninges. The patient was positioned in knee abdominal position in left lateral posture. The lumbosacral or lumbothoracic area (per requirement) was thoroughly cleaned with antiseptic agents followed by draping with a cut sheet. Because the epidural procedure might lead to discomfort, the area was locally anaesthetized with 2% lignocaine and a 26-gauge needle.

Following the method of loss of resistance (LOR), the epidural catheter was introduced and fixed depending on the level of injury. The patency of the epidural catheter was assessed by injecting normal saline. After ascertaining the strength, approximately 0.5–0.7 mL of hESCs were transplanted per vertebral space. This was followed by the multiple dosing of hESCs using the same catheter in situ through an infusion pump at the rate of 60 mL/hour. Because the hESCs possess a shelf life of half an hour, they were introduced in five to 10 minutes. In the case of higher level of injury, that is, high thoracic or cervical injury, the lumbar catheter was replaced by a single-shot cervical epidural injection. Patients were able to receive fluids intravenously, and pulse, blood pressure, and oxygen levels were checked constantly. After the procedure, the catheter was removed from the back, and the patient was asked to lie on the bed until ready to move.

Because tenderness may be felt at the site of needle insertion for a few hours after the implantation, patients were able to receive an ice pack, which could be applied for 10–15 minutes once or twice a day. Patients were asked to rest for the remainder of the day, with normal activities typically being resumed the following day. A temporary increase in pain is possible for several days after due to the pressure of the fluid injected or the local inflammatory response.

The epidural procedure was contraindicated if the patient was on anti-coagulant therapy (which might lead to coagulopathy) or suffering from fever, local sepsis, or local infection.

Caudal Procedure

Administration through the sacral membrane—approximately three centimeters above the tip of the coccyx and in continuum with the epidural space—can be achieved by a 'caudal' injection.

Using the LOR technique, hESCs were transplanted through one of the supplemental routes, the caudal epidural space. The patient was asked to lie in a fetal position, and the underlying sacral hiatus was located via the skin folds of the buttocks. With the tip of the coccyx in the natal cleft, the thumb of the same hand was used to palpate the sacral cornua. The sacral area was cleaned thoroughly with an antiseptic solution and was properly draped. The area was numbed with a local (lignocaine 2%), and a special needle of 26 G (two inches) was introduced via the sacro-coccygeal membrane at an angle of 45 degrees. Positioning is accurate when a distinct 'pop' is heard. The needle was further penetrated parallel to the sacrum, and 5 mL of air was injected with hands positioned over the side of the needle tip. On feeling no air nor tissue resistance, hESCs were introduced with a 26 G needle. A number of complications can arise following a caudal procedure, and the physician must take great care at the insertion site. If the needle has been inserted correctly, it will swing easily from side to side at the hub while the shaft is held like a fulcrum at the sacro-coccygeal membrane, and the tip moves freely in the sacral canal. An early resistance of insertion will show incorrect placement. A caudal injection is not advised if the patient refuses or if there is infection at the site, hypovolemic shock, coagulopathies, or preexisting neurologic disease.

During the treatments, no adverse effects or complications were observed. Clinical as well as radiological improvements were followed by regeneration of the spinal cord (Fig. 6).

A preper anesthesia process focuses on rapid recovery and ensures reduced pain. It also results in few complications and minimal systematic changes during the entire transplantation period. The use of anesthesia during stem cell transplantation in the human body has also been reported in many studies (Sharma et al. 2014; Negrin 2015). Lignocaine is a popularly used local anesthetic for spinal anesthesia and is mainly used for short surgical procedures due to its predictable onset and dense sensory and motor-blocking capacity for moderate duration. Many reports, however, suggest the neuro-toxic effects of lignocaine, thus doubting the use of lignocaine for spinal anesthesia (Schneider et al. 1993; Hampl et al. 1995; Freedman et al. 1998). However, some studies favor the use of

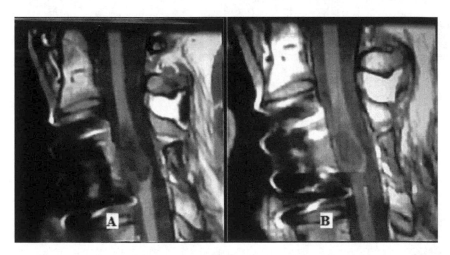

Fig. 6 Tractographic images of a SCI patient before and after hESC therapy using magnetic resonance imaging (MRI)

lignocaine as an excellent and safe modality for patients undergoing surgery (Srivastava et al. 2004). When lignocaine was used as an anesthetic agent prior to the procedure, no neuro-toxic effects were observed.

The anesthesiologist can provide specialized pre- and post-operative medical care (White et al. 2007), train others participating in the procedure, and evaluate patients before and after the procedure, adding depth to the physician-patient relationship. The complexity of the procedure's associated patho-physiology and risks require 'one-on-one' attention from the trained, experienced anesthesiologist. During transplantation, pain control is more effective due to the anesthesiologist-patient relationship developed during various consultations (Yosaitis, Manley, and Plotkin 2005) (Fig. 7).

Patient care and management should be a multidisciplinary strategy rather than a sub-specialty limited to one medical profession. The anes-

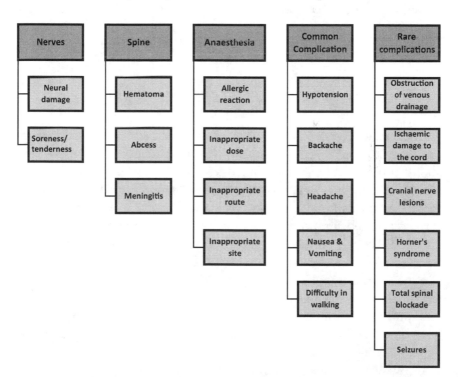

Fig. 7 Risks associated with epidural and caudal route of administration

thesiologist's role would ideally reach beyond the time of hospitalization, as effective pain control and appropriate hESC implantation is as important as achieving successful convalescence.

An interdisciplinary approach applied for the rehabilitation of patients with SCI also includes a team consisting of clinicians skilled in hESC therapy and physiotherapists. The patient data suggests positive role of physiotherapy in improving the mobilization in patients with SCI receiving hESC therapy (Shroff et al. 2016). It reveals an improvement in mobilization in patients with chronic SCI after receiving a combination of hESC and physical therapy. The physical therapy aided in training of cells and took care of atrophy of limbs, whereas hESC therapy resulted in an overall improvement of the patients. This has been observed due to the reduction in the orthotic devices and use of mobility aids. A previous study also showed remarkable improvement in the clinical, locomotive, as well as functional symptoms of the patients, where 81.72 % were able to walk with the support of calipers and mobility aids after receiving hESC therapy (Shorff et al 2015).

Conclusion

The case studies detailed in this chapter are the first of their kind to demonstrate the adequate efficacy of hESC in SCI patients with a good tolerability profile. Our patients gained voluntary movement of the areas below the level of injury as well as improvements in bladder and bowel sensation and control, gait, and hand grip. The MRI and tractography images taken before and after therapy confirmed the improvements observed. We did not observe any difference in the response to therapy between men and women.

Patient improvement was reflected in MRI scans and tractography reports that showed regeneration of the lost axonal connections. Fear of teratomas and immune rejection hinder the use of hESC therapy, but none of the patients in our study had a teratoma or an immune response; patients were not given steroids or immunosuppressants. Adverse events were mild and resolved without any sequel, with headache and fever being the most common. It has been reported that inadvertent dural puncture can lead to the post-dural puncture headache (Crawford 1980). We began using hESCs for treatment in the year 2002, and our first patient received only four doses that year; no adverse events occurred until 2004, and the patient had experienced great benefits (Shroff and Barthakur 2015).

The results of the present study have given a new treatment option to SCI patients. We considered all patients for analyses regardless of the level and extent of injury.

This now-permanent disability affects the everyday lives of those with SCI. Even small clinical improvements can contribute to a better well-being and more productive life. hESC transplantation in SCI patients presents a unique opportunity to address this mostly unmet medical need.

Notes

1. The data in this chapter first appeared in open-access articles distributed under the terms of the Creative Commons Attribution License in the following peer reviewed journals:

   1. Geeta Shroff, Puneet Agarwal, Avinash Mishra, and Nayan Sonowal. 2015. Human Embryonic Stem Cells in Treatment of Spinal Cord Injury: A Prospective Study. *Journal of Neurological Research*, 5(3): 213–220.
   2. Geeta Shroff and Rakesh Gupta. 2015. Human Embryonic Stem Cells in the Treatment of Patients with Spinal Cord Injury. *Annals of Neurosciences*, 22(4): 208–216.
   3. Geeta Shroff, Nayan Sonowal, and Avinash Mishra. 2015. Role of Anesthetists in Human Embryonic Stem Cells Transplantation in Patients with Spinal Cord Injury. *Journal of Anesthesia & Clinical Research*, 6(5): 1–6.
   4. Geeta Shroff. 2016. Human Embryonic Stem Cell Therapy in Chronic Spinal Cord Injury: A Retrospective Study. *Clinical Translational Science*, 00: 1–8.

References

ASIA/IMSOP. 1996. *American Spinal Injury Association/International Medical Society of Paraplegia (ASIA/IMSOP) International Standards for Neurological and Functional Classification of Spinal Cord Injury patients (Revised)*. Chicago, IL: American Spinal Injury Association.

Bergua, F.J.B., J.R.T. Huamán, S.M. Castilla, F.M. Bermudo, B.S. Escoms, et al. n.d. Patent: Method for Differentiation of Pluripotent Stem Cells into Definitive Endoderm Cells.

Bjorklund, L.M., R. Sánchez-Pernaute, S. Chung, T. Andersson, I.Y. Chen, McNaught KS, and A.L. Brownell. 2002. Embryonic Stem Cells Develop into Functional Dopaminergic Neurons after Transplantation in a Parkinson Rat Model. *Proceedings of the National Academy of Sciences of the United States of America* 99: 2344–2349.

Bracken, M.B. 2012. Steroids for Acute Spinal Cord Injury. *Cochrane Database of Systematic Reviews* 1: CD001046.

Bracken, M.B., M.J. Shepard, W.F. Collins, T.R. Holford, W. Young, D.S. Baskin, H.M. Eisenberg, et al. 1990. A Randomized, Controlled Trial of Methylprednisolone or Naloxone in the Treatment of Acute Spinal-cord Injury. Results of the Sec- ond National Acute Spinal Cord Injury Study. *The New England Journal of Medicine* 322 (20): 1405–1411.

Bretzner, F., F. Gilbert, F. Baylis, and R.M. Brownstone. 2011. Target Populations for First-in-Human Embryonic Stem Cell Research in Spinal Cord Injury. *Cell Stem Cell* 8 (5): 468–475.

Crawford, J.S. 1980. Experiences with Epidural Blood Patch. *Anaesthesia* 35: 513–515.

Dalbayrak, S., O. Yaman, and T. Yilmaz. 2015. Current and Future Surgery Strategies for SCI. *World Journal of Orthopedics* 6 (1): 34–41.

Erceg, S., Sergio Laínez, Mohammad Ronaghi, Petra Stojkovic, Maria Amparo Pérez-Aragó, Victoria Moreno-Manzano, Rubén Moreno-Palanques, Rosa Planells-Cases, and Miodrag Stojkovic. 2008. Differentiation of Human Embryonic Stem Cells to Regional Specific Neural Precursors in Chemically Defined Medium Conditions. *PLoS One* 3: e2122.

Freedman, J.M., D.K. Li, K. Drasner, M.C. Jaskela, B. Larsen, and S. Wi. 1998. Transient Neurologic Symptoms after Spinal Anesthesia: An Epidemiologic Study of 1,863 Patients. *Anesthesiology* 89: 633–641.

Grossman, R.G., R.F. Frankowski, K.D. Burau, E.G. Toups, J.W. Crommett, M.M. Johnson, M.G. Fehlings, et al. 2012. Incidence and Severity of Acute

Complications After Spinal Cord Injury. *Journal of Neurosurgery. Spine* 17 (1 Suppl): 119–128.

Gupta, N., J.M. Solomon, and K. Raja. 2008. Demographic Characteristics of Individuals with Paraplegia in India—A Survey. *Indian Journal of Physiotherapy and Occupational Therapy* 2: 24–27.

Hampl, K.F., M.C. Schneider, W. Ummenhofer, and J. Drewe. 1995. Transient Neurologic Symptoms After Spinal Anesthesia. *Anesthesia and Analgesia* 81: 1148–1153.

Kang, K.S., S.W. Kim, Y.H. Oh, J.W. Yu, K.Y. Kim, H.K. Park, C.H. Song, and H. Han. 2005. A 37-year-old Spinal Cord-Injured Female Patient, Transplanted of Multipotent Stem Cells from Human UC Blood, with Improved Sensory Perception and Mobility, Both Functionally and Morphologically: A Case Study. *Cytotherapy* 7: 368–373.

Keirstead, H.S., G. Nistor, G. Bernal, M. Totoiu, F. Cloutier, K. Sharp, and O. Steward. 2005. Human Embryonic Stem Cell-Derived Oligodendrocyte Progenitor Cell Transplants Remyelinate and Restore Locomotion after Spinal Cord Injury. *The Journal of Neuroscience* 25: 4694–4705.

Kerr, D.A., J. Llado, M.J. Shamblott, N.J. Maragakis, D.N. Irani, T.O. Crawford, C. Krishnan, et al. 2003. Human Embryonic Germ Cell Derivatives Facilitate Motor Recovery of Rats with Diffuse Motor Neuron Injury. *The Journal of Neuroscience* 23 (12): 5131–5140.

Kirshblum, S., S. Millis, W. McKinley, and D. Tulsky. 2004. Late Neurologic Recovery after Traumatic Spinal Cord Injury. *Archives of Physical Medicine and Rehabilitation* 85: 1811–1817.

Kirshblum, S.C., S.P. Burns, F. Biering-Sorensen, W. Donovan, D.E. Graves, Amitabh Jha, Mark Johansen, et al. 2011. International Standards for Neurological Classification of Spinal Cord Injury (Revised 2011). *The Journal of Spinal Cord Medicine* 34: 535–546.

Lee, H., G.A. Shamy, Y. Elkabetz, C.M. Schofield, N.L. Harrsion, G. Panagiotakos, N.D. Socci, V. Tabar, and L. Studer. 2007. Directed Differentiation and Transplantation of Human Embryonic Stem Cell-derived Motoneurons. *Stem Cells* 25: 1931–1939.

Leuty, R. 2014. Stem Cell Trial for Spinal Cord Injuries Cleared by FDA. http://www.bizjournals.com/sanfrancisco/blog/biotech/2014/08/embryonic-stem-cells-asterias- geron-spinal-cord.html. Accessed 17 September 2014.

Li, Y.W., L. Ma, B. Sui, C.H. Cao, and X.D. Liu. 2014. Etomidate with or Without Flumazenil Anesthesia for Stem Cell Transplantation in Autistic Children. *Drug Metabolism and Drug Interactions* 29: 47–51.

Lima, C., P. Escada, J. Pratas-Vital, C. Branco, C.A. Arcangeli, G. Lazzeri, C.A. Maia, et al. 2010. Olfactory Mucosal Autografts and Rehabilitation for Chronic Traumatic Spinal Cord Injury. *Neurorehabilitation and Neural Repair* 24: 10–22.

Lukovic, D., V. Moreno Manzano, M. Stojkovic, S.S. Bhattacharya, and S. Erceg. 2012. Concise Review: Human Pluripotent Stem Cells in the Treatment of Spinal Cord Injury. *Stem Cells* 30: 1787–1792.

Lukovic, D., M. Stojkovic, V. Moreno-Manzano, S.S. Bhattacharya, and S. Erceg. 2014. Perspectives and Future Directions of Human Pluripotent Stem Cell-based Therapies: Lessons from Geron's Clinical Trial for Spinal Cord Injury. *Stem Cells and Development* 23: 1–4.

Mackay-Sim, A., F. Feron, J. Cochrane, L. Bassingthwaighte, C. Bayliss, W. Davies, P. Fronek, et al. 2008. Autologous Olfactory Ensheathing Cell Transplantation in Human Paraplegia: A 3-year Clinical Trial. *Brain* 131 (Pt 9): 2376–2386.

McDonald, J.W., X.Z. Liu, Y. Qu, S. Liu, S.K. Mickey, D. Turetsky, D.I. Gottlieb, and D.W. Choi. 1999. Transplanted Embryonic Stem Cells Survive, Differentiate and Promote Recovery in Injured Rat Spinal Cord. *Nature Medicine* 5: 1410–1412.

Negrin, R.S. 2015. Patient Information: Bone Marrow Transplantation (Stem Cell Transplantation). Beyond the Basics.

Paralysis, Paraplegia, and Quadriplegia. MD guidelines. 2015. http://www.mdguidelines.com/paralysis-paraplegia-and-quadriplegia. Accessed 24 December 2015.

Park, H.C., Y.S. Shim, Y. Ha, S.H. Yoon, S.R. Park, B.H. Choi, and H.S. Park. 2005. Treatment of Complete Spinal Cord Injury Patients by Autologous Bone Marrow Cell Transplantation and Administration of Granulocyte-Macrophage Colony Stimulating Factor. *Tissue Engineering* 11: 913–922.

Rahimi-Movaghar, V., M.K. Sayyah, H. Akbari, R. Khorramirouz, M.R. Rasouli, M. Moradi-Lakeh, F. Shokraneh, and A.R. Vaccaro. 2013. Epidemiology of Traumatic Spinal Cord Injury in Developing Countries: A Systematic Review. *Neuroepidemiology* 41: 65–85.

Ronaghi, M., S. Erceg, V. Moreno-Manzano, and M. Stojkovic. 2010. Challenges of Stem Cell Therapy for Spinal Cord Injury: Human Embryonic Stem Cells, Endogenous Neural Stem Cells, or Induced Pluripotent Stem Cells? *Stem Cells* 28: 93–99.

Rossi, S.L., G. Nistor, T. Wyatt, H.Z. Yin, A.J. Poole, J.H. Weiss, M.J. Gardener, et al. 2010. Histological and Functional Benefit Following Transplantation of

Motor Neuron Progenitors to the Injured Rat Spinal Cord. *PLoS One* 5 (7): e11852.

Schneider, M., T. Ettlin, M. Kaufmann, P. Schumacher, A. Urwyler, K. Hampl, and A. von Hochstetter. 1993. Transient Neurologic Toxicity After Hyperbaric Subarachnoid Anesthesia with 5% Lidocaine. *Anesthesia and Analgesia* 76: 1154–1157.

Sharma, A., H. Sane, N. Gokulchandran, P. Badhe, P. Kulkarni, and A. Paranjape. 2014. Stem Cell Therapy for Cerebral Palsy – A Novel Option. In *Cerebral Palsy – Challenges for the Future*. Rijeka, Croatia: InTech. 32.

Sharp, J., J. Frame, M. Siegenthaler, G. Nistor, and H.S. Keirstead. 2010. Human Embryonic Stem Cell-Derived Oligodendrocyte Progenitor Cell Transplants Improve Recovery After Cervical Spinal Cord Injury. *Stem Cells* 28 (1): 152–163.

Shroff, G. 2005. *Human Embronic Stem Cells—A Revolution in Therapeutics*. New Delhi: NuTech Mediworld. isbn:81-7525-660-5.

———. 2015a. Establishment and Characterization of a Neuronal Cell Line Derived from a 2-Cell Stage Human Embryo: Clinically Tested Cell-based Therapy for Neurological Disorders. *International Journal of Recent Scientific Research* 6 (4): 3730–3738.

———. 2015b. A Novel Approach of Human Embryonic Stem Cells Therapy in Treatment of Friedreich's Ataxia. *International Journal of Case Reports and Images* 6 (5): 261–266.

———. 2015c. Human Embryonic Stem Cells in the Treatment of Spinocerebellar Ataxia: A Case Series. *Journal of Clinical Case Reports* 4: 474.

———. 2015d. Treatment of Lyme Disease with Human Embryonic Stem Cells: A Case Series. *Journal of Neuroinfectious Diseases* 6: 167.

Shroff, G., and J.K. Barthakur. 2015. Safety of Human Embryonic Stem Cells in Patients with Terminal/Incurable Conditions—A Retrospective Analysis. *Annals of Neurosciences* 22: 132–138.

Shroff, G., and L. Das. 2014. Human Embryonic Stem Cell Therapy in Cerebral Palsy Children with Cortical Visual Impairment: A Case Series of 40, Patients. *Journal of Cell Science and Therapy* 5: 189.

Shroff, G., and R. Gupta. 2015. Human Embryonic Stem Cells in the Treatment of Patients with Spinal Cord Injury. *Annals of Neurosciences* 22 (4): 208–216.

Shroff, G., A. Gupta, and J.K. Barthakur. 2014. Therapeutic Potential of Human Embryonic Stem Cell Transplantation in Patients with Cerebral Palsy. *Journal of Translational Medicine* 12: 318.

Shroff, G., Dipin Thakur, Varun Dhingra, Deepak Singh Baroli, Deepanshu Khatri, and Rahul Dev Gautam. 2016. Role of Physiotherapy in the Mobilization of Patients with Spinal Cord Injury Undergoing Human Embryonic Stem Cells Transplantation. *Clinical and Translational Medicine 5 (41): 1–9*.

Srivastava, U., A. Kumar, S. Saxena, R. Saxena, N.K. Gandhi, and P. Salar. 2004. Spinal Anaesthesia with Lignocaine and Fentanyl. *Indian Journal of Anaesthesia* 48: 121–123.

Ware, C.B., A.M. Nelson, B. Mecham, J. Hesson, W. Zhou, E.C. Jonlin, A.J. Jimenez-Caliani, et al. 2014. Derivation of Naive Human Embryonic Stem Cells. *Proceedings of the National Academy of Sciences of the United States of America* 111: 4484–4489.

White, P.F., H. Kehlet, J.M. Neal, T. Schricker, D.B. Carr, F. Carli, and Fast-Track Surgery Study Group. 2007. The Role of the Anesthesiologist in Fast-track Surgery: From Ultimodal Analgesia to Perioperative Medical Care. *Anesthesia and Analgesia* 104: 1380–1396.

Willerth, S.M., and S.E. Sakiyama-Elbert. 2008. Cell Therapy for Spinal Cord Regeneration. *Advanced Drug Delivery Reviews* 60: 263–276. Epidural: The Indications and Contraindications for Epidural Nalgesia.

World Medical Association Declaration of Helsinki.

Yoon, S.H., Y.S. Shim, Y.H. Park, J.K. Chung, J.H. Nam, M.O. Kim, H.C. Park, et al. 2007. Complete Spinal Cord Injury Treatment Using Autologous Bone Marrow Cell Transplantation and Bone Marrow Stimulation with Granulocyte Macrophagecolony Stimulating Factor: Phase I/II Clinical Trial. *Stem Cells* 25 (8): 2066–2073.

Yosaitis, J., J. Manley, L. Johnson, and J. Plotkin. 2005. The Role of the Anesthesiologist as an Integral Member of the Transplant Team. *HPB: The Official Journal of the International Hepato Pancreato Biliary Association* 7: 180–182.

Geeta Shroff is Founder and Director of Nutech Mediworld, New Delhi, India. Dr. Shroff has developed the technology to isolate human embryonic stem cells (hESC), culture them, prepare them for clinical application, and store them in ready-to-use form with a shelf life of six months. Further, this technology is being used clinically to treat patients. Since 2002, more than 1400 patients suffering from various conditions categorized as incurable—spinal-cord injury, diabetes, multiple sclerosis, Parkinson's disease, cardiac conditions, cerebral palsy—have been treated successfully by Dr. Shroff, and the number is steadily growing. A graduate in Medicine from the University of Delhi, Dr. Shroff did her postgraduation work in Gynecology and Obstetrics. She specialized in treating infertility and is a trained embryologist and a qualified IVF practitioner. After eight years of valuable clinical experience at Safdarjung and Batra hospitals, large multi-specialty hospitals in Delhi, Dr. Shroff set up her own IVF practice in 1996. She began research on human embryonic stem cells in 1999 and pioneered hESC therapy. She has presented her work at various national and international forums. Dr. Shroff envisions making hESC therapy available globally so that it becomes the first line of treatment for many of humankind's worst afflictions.

Pre-blastomeric Regeneration: German Patients Encounter Human Embryonic Stem Cells in India

Petra Hopf-Seidel

As a psychiatrist and neurologist in a small, private medical office in Germany, I specialize in treating patients with chronic neurological diseases like Lyme disease and other progressive neurodegenerative disorders like amyotrophic lateral sclerosis (ALS), multiple sclerosis, multiple chemical sensitivity, or chronic fatigue syndrome.

My first encounter with this newly developed form of stem cell therapy was at a Lyme disease conference in Saarbrücken in the spring of 2012. The organizer was a 39-year-old female patient of mine who had been suffering for the last 7 years from a debilitating form of Lyme with multiple sclerosis-like symptoms: she had exhaustive fatigue and weakness as well as paralysis of both legs, which had left her wheelchair bound. In addition, she had cognitive impairments, bilateral neuritis of the optic nerve and glaucoma, and intense musculoskeletal pains. The last time I had seen her in my medical office, some months earlier, she had been very ill.

To my great surprise, she was at the conference running about on high heels, managing this large event. She had just returned from her second

P. Hopf-Seidel (✉)
Ansbach, Germany

© The Author(s) 2018
A. Bharadwaj (ed.), *Global Perspectives on Stem Cell Technologies*,
https://doi.org/10.1007/978-3-319-63787-7_6

trip to Nutech Mediworld in New Delhi, and at this conference, I met Dr. Geeta Shroff and heard her lecture on human embryonic stem cell (hESC) therapy and the possibilities of these specific cell lines (see chapter 'Establishment and Use of Injectable Human Embryonic Stem Cells for Clinical Application'). After seeing astonishing pictures of how traumatic spinal cord injuries could be mended with hESC therapy, I decided to travel to India with some chronically ill patients who were desperate because no conventional treatment had helped them. I wanted to learn about the type of treatment they would receive, and I wanted to help my patients with language problems and assist them in adapting to a foreign country. Also, I wanted to understand the specific properties of these stem cell types currently under scientific investigation.

A Chance to Jump

I chose three female patients between the ages of 20 and 57 who had been suffering for years with Lyme neuroborreliosis as a consequence of a *Borrelia burgdorferi* infection many years ago. Before departing for India, all three were treated thoroughly with antibiotics; none had signs of Lyme at the time of travel. But they had neurological damage resulting from the long-lasting spirochete infection, namely, paralysis, cognitive impairment, exhaustion, fatigue, and gastrointestinal issues.

Patient 1 was 48 years old at the time of treatment and had originally been diagnosed with multiple sclerosis, despite the onset of her symptoms beginning after a tick bite with an erythema migrans. Her MRI had shown lesions in her cerebral white matter and spinal cord, which were unfortunately incorrectly only diagnosed as an indication of multiple sclerosis without taking in consideration the possibility of a Borrelia burgdorferi infection as the cause for the cerebral lesions. When she began coming to my office, she had been undergoing 17 years of immunosuppressant treatment for her supposed multiple sclerosis; she was also spastic paraplegic and had slurred speech and a great deal of cognitive problems. She was also wheelchair bound.

Patient 2 was a 57-year-old woman who had not been diagnosed with Lyme neuroborreliosis despite a documented tick bite 6 years ago and many MRI-documented lesions in her cervical spinal cord. Due to a spastic condition in her right leg, she was unable to walk and was in a wheelchair.

Patient 3 was a 20-year-old young woman with weak muscles but was still able to walk with support. She suffered from exhaustion, was malnourished (she arrived in India with a body weight of 41 kg [BMI 14]), and had severe cognitive problems. She had been suffering from Lyme neuroborreliosis for 5 years and had experienced three episodes of temporary tetraparesis caused by Guillain–Barré syndrome. Although she was treated successfully with intravenous immunoglobulins, the severe muscle weakness remained as well as many gastrointestinal and cognitive problems.

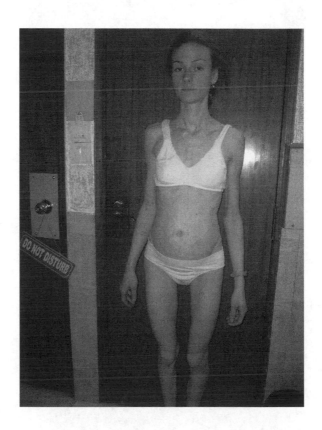

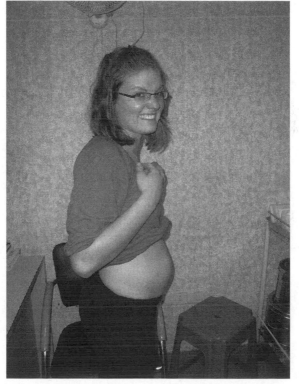

During their 8-week stay, they had daily hESC injections, which were administered intramuscularly, intravenously, and even intrathecally (in the operation theatre, under sterile conditions) as well as intranasally and orally. They underwent physiotherapy twice a day and followed a protocol of vitamins, minerals, and proteins. Fluoxetine was given as a support for new cells to grow (as it was microscopically observed to happen) and minocycline, an antibiotic, was given as protection against superinfection. No adverse effects were reported.

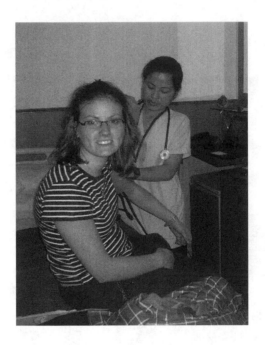

Unfortunately, during treatment, Patient 1 broke her thigh when she fell out of her wheelchair; this prevented her from exercising. Although the hESC treatment did not improve her spastic paralysis, her slurred speech showed improvement. However, after 1 week of treatment, Patient 2 was able to walk slowly with a cane inside the Taj Mahal, where no wheelchairs are allowed. By the end of their stay,

Patients 1 and 2 were still in wheelchairs and the spasticity in their legs had not improved. However, Patient 1, with the broken thigh, had the chance to return a few months later, after her leg had healed. Patient 2 did not feel her walking had improved even though physiotherapists thought she had progressed.

Patient 3 showed the greatest success. Both her brain single photon emission computed tomography (SPECT) scans show an amazing improvement. At the beginning of her stay her SPECT shows blue and black brain areas, meaning little or no circulation. However, her February 21, 2013, SPECT scan shows nearly normal (pink) blood circulation. In addition, after only 1 week of treatment, the patient who was unable to walk unaided was able to jump. She also experienced fewer gastrointestinal issues, had a weight gain of 7 kg and greater muscle strength, and had much better cognitive abilities.

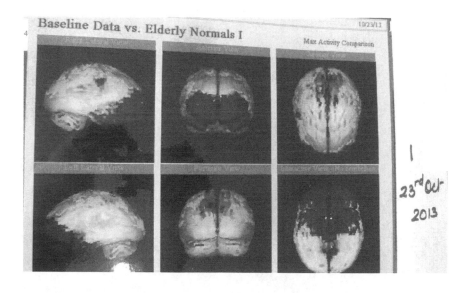

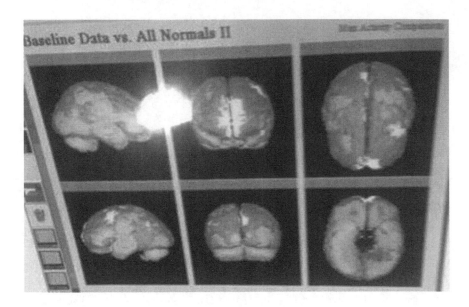

Sightseeing and Strolling

Encouraged by the potential of hESC, I went again to India with six of my chronically and seriously ill patients.

Patient 3 from my first trip wanted a second treatment because some of her previous problems had begun to reappear (e.g., nausea, loss of appetite, extreme bloating, cognitive impairments, and some muscle weakness). During this 8-week round, she had physiotherapy twice a day, and her hESC was given through many application routes (including oral, nasal, intramuscular, and intravenous). Her gastrointestinal problems ceased, she regained some weight and muscle strength, and she was able to shop and sightsee in New Delhi, activities she had not been able to do for the past 6 years. She left India very happy and much healthier than when she came, especially in comparison to her first visit only 10 months before.

Patient 4 was a female Lyme neuroborreliosis patient who had spastic paraparesis after an insect bite (most likely a spider). The reason for her neurological problems was an infection with *Bartonella henselae* and *Borrelia burgdorferi* at the age of 11 with an immediate paralysis of her legs afterwards. By the time of treatment in India, she had been in a wheelchair for 9 years. She had a flabby paralysis of her left arm and was prone to infections, especially virus reactivations, and had severe cognitive deficits. Additionally, she suffered vegetative problems like extreme sweating and sleeplessness with change of the day–night pattern. Her case was very challenging, as the flabby paralysis of her left arm and spastic paraparesis had been her condition for 9 years. However, her left hand responded to the treatment with some movements and new spasticity with extension of her fingers, which helped her use the hand in special positions. Although this condition only lasted a few months and her previous condition has returned, her cognitive abilities, alertness, and stamina are much better, and her SPECT scans show this result. She was able to sit upright with less spasticity in her back. She also underwent a second round of treatment for 4 weeks, which nevertheless helped her sweat less and sleep better. She was also less prone to infections and viral reactivations.

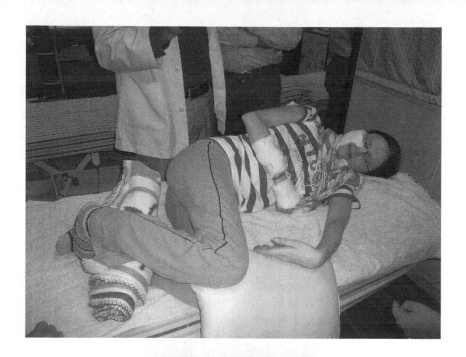

Patient 5 was a male, a medical surgeon by profession, who had had ALS for the past 4 years. He was very weak and had gastrointestinal issues and a resulting loss of weight, respiratory problems, a low and hoarse voice, and recurrent atrial tachycardias. He was also infected with *Borrelia burgdorferi* and had many signs of chronic Lyme. Additionally, he was suffering of a mercury load which caused a Type IV allergy. This and the Lyme activity were treated thoroughly before leaving for India. He received hESC through every possible route: intrathecal in the sterile operation theatre, intramuscular, intravenous, oral, and nasal. After 1 week, he was able to move in an upright position after many weeks of stooping. His overall condition improved greatly, and at the end of his stay he was even able to climb stairs. He could eat and speak normally and had no respiratory problems. He planned a second round for the end of January 2014 to keep up his improvements. After his fourth round and with oxygen support at night, he amazingly could work 4 hours a day.

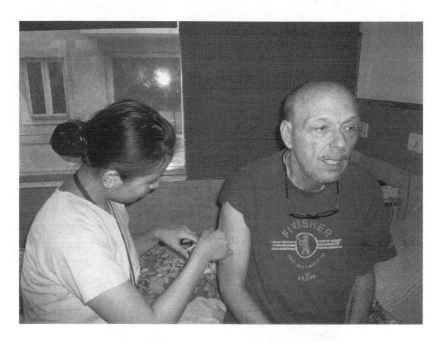

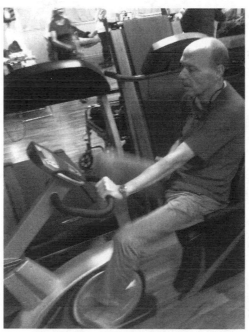

Patient 6 was a 65-year-old female with a more advanced stage of ALS. My diagnosis was that she was also suffering from chronic Lyme and had a burden of mercury. Both these conditions were treated thoroughly before leaving for India. This was her fifth treatment: although she had recovered well after each hESC treatment, her condition always deteriorated. Before this trip, she could not hold her head upright, was in a wheelchair, and had weakness in both arms and hands. After her treatment, she was able to hold her head up, was less depressed, and had a stronger grip in both hands. Although her muscle weakness from ALS was already quite advanced and she was not able to stand on her own, after this treatment, she was able to sit without support.

Patient 7 was a 73-year-old female with a family history of a dry macular degeneration. Her left eye had a reduced vision of 30 per cent, and her right eye showed wet macular degeneration. Her treatment was planned for only 2-and-a-half weeks. Accompanying her was Patient 8, her husband, a 74-year-old with arthritis in both knees. He was unable to bend them without pain, let alone climb stairs. His treatment was to be given intramuscularly but mostly locally on both knees. His treatment, too, was planned for 2-and-a-half weeks. After this time, he went sightseeing and could walk long distances; he could even climb stairs without any pain. His wife's vision improved from 30 to 40 per cent during treatment and, back home, to 60 per cent.

I must say that each of my patients improved in one way or other, some visible and measurable, others more invisible with improved stamina, better moods, or more muscle strength. No one had adverse effects, so I can say confidently that I was surrounded by happy patients.

Walking and Studying

In March 2015, I decided to return to India with three patients, with a fourth patient joining us in New Delhi.

Patient 9 was a 51-year-old man with chronic Lyme and a burden of cobalt and nickel combined with a genetic inability to excrete the metals. He had also, for the past 2 years, been showing signs of ALS. This was his third hESC treatment and was planned for 5 weeks. However, in the

3 months before the trip, his muscle strength and overall abilities had deteriorated quickly, and he was now showing all the clinical symptoms of ALS. He was in a wheelchair, and his SPECT showed a severe deficiency in blood circulation.

Unfortunately, he could not be helped. His weakness worsened, and he was still in his wheelchair. He also had difficulty swallowing, lost weight, and could only speak in a low voice. Back home, the situation deteriorated further, and, by the end of 2015, he was bedridden and could not move his limbs and was only able to 'speak' with his eyes. His ALS progressed rapidly; all three phases of the hESC therapy were not able to slow it down.

Two of the patients on this visit had more positive results. Patients 10 and 11 were a 71-year-old female 73-year-old male, both suffering from arthritis in both knees. They were scheduled for 2 weeks of treatment administered only locally and intramuscularly and after treatment could walk without pain, a condition that still remains.

Patient 12 was a 36-year-old law student who came for his second round of treatment because of extreme exhaustion and weakness (diagnosed as chronic fatigue syndrome) as well as severe memory loss induced by chronic Lyme and due to a re-infection with Borrelia burgdorferi in December 2003. Furthermore, he had a burden of formaldehyde and a genetic inability of excretion because of some polymorphisms in his glutathione S-transferase genes. His January 2014 brain SPECT in Germany showed a hypoperfusion of the left frontal and temporal lobe, explaining well his 'brain fog' and loss of short-term memory. After this treatment round, he experienced the same positive results as after his first. His alertness and stamina improved, as did his cognitive abilities with better short-term memory. He was able to study again for several hours a day for his last law examination before graduation.

Summary of Patient Results

Overall, I had accompanied 12 patients (7 female and 5 male) between the ages of 20 and 73. One of my patients, a woman, went to India by herself in 2010 and 2011 and experienced great improvements in her physical and mental conditions.

Three of the patients I accompanied were in different stages of ALS. Two improved temporarily but relapsed each time they returned home. This is not surprising, as ALS is a very progressive disease with, until now, no effective treatment. The female ALS patient underwent five treatments but died in December 2015. The surgeon, Patient 5, has so far had four treatment protocols and is able to walk with a walker or sit in his wheelchair. The youngest of my three ALS patients had such an aggressive form of the disease that he is now bedridden and paralysed and seems to be in the terminal phase of this terrible disease.

Of the six patients suffering from the consequences of severe and chronic Lyme, four of them had spastic paralysis of their legs, and three were in wheelchairs. The two youngest female patients were very malnourished, with a very low body mass index because of their long-lasting gastrointestinal issues associated with the disease. Spastic paralysis did not resolve for any of these patients, but their overall strength and stamina improved as well as their 'brain fog' and vegetative disorders. Their gastrointestinal problems greatly improved. One gained 16 kg after her second round of treatment and now lives the life of a healthy young woman, with no flatulence or poor digestion and no muscle weakness.

The best results were from the arthritis and macular degeneration patients, all of whom had treatments lasting only 2-and-a-half weeks and were otherwise healthy. After their treatments, they could walk and climb stairs without any pain, and their success did not wane over time. But for such relatively smaller complaints, I would think nobody would travel to India.

The most severely ill patients, with long-standing impairments like spastic paraparesis or with ALS, could not be helped as much as hoped. They would have to keep returning to India several times in certain intervals to sustain their improvements. Therefore, what would help them and other severely ill patients most would be hESC therapy in Europe and other countries. If this effective treatment were available on a worldwide basis, then many of our currently incurable diseases and conditions could be improved upon or even healed. I fully agree with Dr. Shroff's vision to see human embryonic stem cells as the first line of treatment for many of mankind's worst afflictions. And I think this holds great promise for the future.

Petra Hopf-Seidel is a neurologist and psychiatrist in Germany. In 1970, Dr. Hopf-Seidel finished the humanistic gymnasium in Bamberg and started professional training as a scientific librarian. She finished the training in 1973 and worked in the Bavarian state library in Munich. In 1974, she began studying medicine in Würzburg and Berlin and graduated in 1979 from the Free University of Berlin. After finishing her thesis (*magna cum laude*), she worked as a surgical-assistant doctor. One year later, she left with her family to live in Malaysia for three and a half years. Following this, she began her postgraduate studies in a psychiatric hospital and trained to become a specialist in family medicine, in neurology and psychiatry. After finishing her postgraduate studies, she worked for two more years in a psychiatric hospital before setting up her medical office as a specialist in neurology and psychiatry. Since 2003, she has had a private practice in neurology and psychiatry, predominantly treating chronically ill patients, most of whom suffer from Lyme disease.

Part III

Patient Positions

Active Parents, Parental Activism: The Adipose Stem Cell In Vitro Lab Study

Ripudaman Singh

In June 2005, my wife and son and I were preparing for our first trip to the United Kingdom, and everything was in order: tickets, visas, foreign exchange. Ten days before our trip, my wife and 4-year-old son travelled to Chandigarh, where both of our parents live, to say goodbye to her parents as well as do some last-minute shopping. I stayed in Delhi to work and save up leave for our trip. Excitement was high.

However, while my wife and son were in Chandigarh, my father-in-law called. My father-in-law is a paediatric doctor, and my mother-in-law is a general practitioner. My son had been diagnosed with some sort of muscle-wasting disease. 'Something doesn't look right', he said. 'It's his calves. They look much bigger than those of a normal child. It's called hypertrophy'.

At that exact moment, I knew our trip abroad would go for a six. After the initial disappointment of cancelling our holiday and the tumult of emotions when I realized my son had a serious 'condition', I began to try

R. Singh (✉)
New Delhi, India

© The Author(s) 2018
A. Bharadwaj (ed.), *Global Perspectives on Stem Cell Technologies*,
https://doi.org/10.1007/978-3-319-63787-7_7

to put the pieces together. The 'condition', as it turned out, was something I'd never heard of, Duchenne muscular dystrophy (DMD). This rare disease is an X-linked chromosome and is autosomal recessive, thus affecting boys at least 99 per cent of the time. For some the life span is 17 years, others 20, and others about 25. One can't be sure. Thus began my initial foray and subsequent deep immersion into a new world of medicine and treatments that I continue to this day. In that initial 4 months, I studied DMD until I understood it as thoroughly as a layman could and found it is a rare, genetic muscle-wasting condition that affects only boys and can lead to early death.

So here I was, a conventional man with a regular middle-class life faced with a cataclysmic challenge concerning my own son. No one in my family had really had a medical problem, certainly not one this big. I was a sensitive, emotionally vulnerable person at the time and was going under with all the tension and strain as well as the painful reality my son may not live that long. I was all but a nervous wreck. The blow turned all my thoughts, ideologies, and presuppositions 180 degrees.

While researching DMD, I still had to work and get on with my life. Every spare minute I was at my computer, deep into the internet, researching, exploring, and networking. It was then that I chanced upon the science of stem cells, which I hadn't heard about until then. It was the first time in my life I had had to make a very serious decision. I had been reading everything possible on stem cells, but I still needed to know they worked. Should I subject my son, a mere child, to a relatively less understood line of treatment? All I knew was that conventional biomedical avenues had nothing to offer. I started talking extensively with other DMD parents, but none had chosen the stem cell option. And then there was a big consideration—the expenses. It was not a decision we could make with a snap of our fingers, especially while we were groping in the dark about DMD and its response to stem cell therapy.

The internet was my life jacket. I am not very technically savvy, but I spent most of my non-office hours scouring the net and connecting with people all over the world who had anything to do with stem cells: researchers, doctors, clinicians, and parents. But when I asked parents if they were doing research on stem cells, they would say they thought the cure for

DMD was still decades away or maybe another 5 years. However, I began to get a hang of stem cell science and to talk to many DMD parents. Almost every city has a DMD support group: Delhi had one and so did Mumbai. I joined the Delhi group and attended a few meetings, but apart from drinking tea and chatting, there was not much action. During these meetings, people would share their children's experiences. If someone said, 'My son fell down and broke his arm', I would become alarmed and depressed. I began to pray regularly, a certain kind of prayer I did every day without fail. I felt this would see me through; after all, the doctors were shutting the doors in our faces. Doctors are not gods, something will happen, a miracle, I would tell myself. These were doctors who were waiting for research journals to land on their tables. Their daily practices kept them busy and padded their wallets. So I stopped going to doctors. But I needed to blame someone for this calamity. So I began to blame God, and believe it or not, that helped me. I would chat with God regularly and spent a lot of time praying.

Four years went by. By now I'd come to the conclusion that I could not wait for things to happen by themselves. I had to gather myself and *do* something. My son's condition was deteriorating; I didn't have the luxury of time. Until he was 8, he was like a normal kid, running around and playing with friends. In DMD, there is what you call a threshold level. I thought he would walk until he was about 13 or 14, but in 2010, when he was 9 and a half, he suddenly announced he couldn't walk. That day I was badly hit, yet again.

I said to myself that praying alone wouldn't solve the issue. I had to take very practical steps. I would buy time if that's was what I needed to do. I got in touch with a researcher who had recently moved to Hyderabad, South India, from Canada, where she had been doing stem cell research. I chatted with her for nearly an hour, and it was a wonderful conversation, just like you would want a researcher or doctor to react towards you. I had by now become a sort of lay expert in DMD and stem cells. I shared a lot of my information with her and she was happy to be informed about what was happening globally regarding stem cells, in particular, for DMD. Around this time, I connected with the Mumbai DMD group, which agreed with my ideas on stem cells, and I had become the 'action man' willing to go the extra mile.

I decided that I would put my son through an experimental clinical trial. It was risky and something that the insurance companies wouldn't touch because it is a genetic disorder with unproven therapies. Unlike other trials, we would need to pay the high cost. I connected four or five fathers from the DMD group with the researcher in Hyderabad, and we told her we would like to be partners in the trial. I couldn't have done it alone. We had also found a paper written by a Brazilian researcher on positive results from adipose stem cell research, and we began to communicate with her, receiving positive feedback on her research. Our goal was to use this research and get an in-vitro cell culture done in the lab. However, even if one lab completes a successful in-vitro study, this does not mean the same can be done successfully elsewhere. The same has to be achieved by another lab, which tries to replicate the study and take it further. Just because it has been done successfully in *some* lab doesn't mean one can start injecting people with it. The same study has to be replicated in *one's own* lab as well.

Although the researcher in Hyderabad was concerned about the cost and the experiment's uncertainties, she agreed to do it with adipose (fat tissue) from a donor who had had liposuction. No animal models were used, because the researcher in Brazil had already done so and to good effect. We spoke to several institutes in India already doing adipose infusions and becoming convinced about their safety. We DMD fathers funded the study, which took 9–12 months, showing good results. The studies showed traces of dystrophin, the protein missing in DMD. We made the brave decision to use adipose-derived mesenchymal stem cell infusions for 14 of our children.

Based on our consultations with the Brazilian researcher, we initially wanted to inject 200 million cells. However, people worldwide were giving 50 million. Then there was a line of treatment that, depending upon the patient's weight per kg, you could give 2 million per kg. So if the child is 40 kg, then you multiply that weight by 2, meaning you would give him 80 million cells. So this was a thought, a scientific thought maybe, but it had no basis as far as I was concerned. Because of our research and discussions with doctors, we felt more cells needed to be given within a specific period of time. Thus, keeping safety in mind, the first infusion was 200 million per child, but we divided those 200 million over 4 infusions or 50 million per week for 4 weeks. Then we came across parents with children who had different conditions, and they were giving 80 million in one shot, and we deduced that if 80 million didn't have adverse reactions, then we could split the 200 million into 100 and 100. Thus, the second infusion was after seven to 8 months with 250 million cells, 125 each. These two infusions went very well, so we stuck to that. About 20–30 per cent of cells are flushed from the body, so cells need to be replenished and infused on a regular basis.

We parents did most of the research and determined the protocols, such as how many cells we wanted and the number of infusions. From our research, we also understood that the blood circulating inside the body at the time the veins are infused is circulating at a temperature of around four degrees centigrade. That inside temperature is what is known, technically, as the hypoxic effect. We asked the researcher in Hyderabad to culture the cells in what is known as a hypoxic chamber so that the cells would be cultured in the same atmosphere found inside the body. When the cells are taken from one atmosphere and injected into a similar one, the cells will have fewer adjustments and the genes won't be shocked.

Of the 14 children who went through the same process, the younger ones showed marked improvement. My son had two rounds of infusions over a 7- to 8-month period, and since he was in a wheelchair, it was difficult to gauge the level of improvement. Until about the age of 9, he seemed like a regular kid. But if one observed him carefully, one could see that he walked on his toes, had an uneven walking pace, and had to put in major efforts to climb stairs. After the treatments, his progress was slow but stable. In other words, whatever was going in was fighting the disease. What I was doing, in effect, was buying time until something better came along.

Meanwhile, our support systems were our wives, who were and are extremely supportive while we did the research and treatment procedures. My wife is a rock star: she handles things much better than I do. Being a daughter of doctors, she was constantly pushing me to do medical things. I don't know where she gets her energy. In all those gruelling years, she never went under, and the last 12 years wouldn't have been possible without her. She was a mountain of strength to me when I was emotionally wrung out. My son is a super star: he is also very positive, has adjusted to his situation, attends school in his wheelchair, and behaves like any normal child. For our part, we have never made him feel that he lacks anything. For us he is perfect.

The usual line from doctors is, 'We have nothing on it right now, we'll see when it happens'. For them it's business as usual. No one tells you anything. I had to educate myself from scratch. If I go to a doctor and say to him, 'Doctor, we want to try stem cell treatment for DMD, he will say, No, don't do it. There is no cure. There is no breakthrough data as of now'. Maybe they were right from their own perspective, but I was looking for hope, involvement, and empathy. But just because he says it, why should I not try it? Does it mean we should pooh pooh the experiments that are happening now? I don't want to go to doctors anymore. More than anything else, I face so much apathy and opposition.

Because the All India Institute of Medical Science (AIIMS) is a government hospital and not looking for profits, they are not as concerned with being leaders in research and wanting to achieve new objectives in this science. They are bogged down with bureaucracy and the typical workings of an Indian government agency; it takes forever to take any decisions. The main advantage of working with them is the possibility of a clinical trial, which could be an effective treatment. I got in touch with AIIMS in 2006, and the head of the stem cell department and the head of paediatrics know me by name. I told my DMD group I was talking to AIIMS about the adipose method, but I was very frustrated by the lack of progress. Initially the group was excited because it was free of cost and AIIMS has the best technology in India. But after 2 years of not making any progress, I am not depending on them anymore. One of the dads in the group is now actively in communication with them, but if they agreed to take this forward, I would rather go with them than anyone else.

For the longest time, I have been interacting with the Indian Council of Medical Research. When I was first trying to understand stem cell science, they were the only ones who could provide information about bona fide stem cell researchers. But when I started asking how they were expediting progress in stem cell research, it was a dead end. They are just sitting on their official seats to give permission for this and that; for example, if someone wanted to put up a stem cell lab, they would okay it. When I was in contact with them, all they would do was criticise people doing stem cell therapies, as if it is their job; for instance, they simply dismiss it, saying stem cells are just placebos. If I asked them to address the problem, they would say they were not the regulators. Who were the regulators? Nobody seemed to have an idea!

I may be harsh when it comes to dealing with the doctors and the government agencies. I do understand their helplessness, as they too are dependent upon our system, which is very frustrating on all levels.

These treatments are expensive. The cells, which now come from Hyderabad, are expensive, and then, of course, there are the hospital charges and the preparations before the infusion therapy, such as blood tests and so forth. Each infusion costs me Rs. 3.5 lakhs once every 6 months, which is a lot of money. Since it's still in an experimental stage, we are getting the treatment at cost, but costs may go up dramatically once stem cells become an accepted line of treatment in a hospital or clinic. Even now, the hospitals in Delhi are making money out of me because they are aware of the situation and helplessness of people like me who are desperate and will try an unproven therapy.

People have been asking me, jocularly, why adipose should be so expensive. Aren't people dying to lose their fat? But we wanted it to be from a woman between the ages of 16 and 21, and many at that age are not fat—at least, they aren't coming for liposuction, and if they don't, it's also because it's expensive. The Hyderabad researcher does manage to find donors. Although those donors don't really mind what use it's put to, we inform them for ethical reasons. In fact, I would say that it's ethical for the plastic surgeon to donate this, especially, since we are not making any profit out of it and are using it on our own kids, but then we are not in a position to bargain.

In Delhi, where I live, the response towards this line of treatment is lukewarm. Many of the DMD parents in this city have a block against trying something new, and when I tell them about the money involved, they think I'm making a sales pitch. Thus, I've stopped trying to discuss or debate it with them, and I just give them the information if they need it. Initially, my in-laws, who are in their 70s and pretty traditional, had no clue about stem cells even though both of them are doctors. But now the entire family—including my parents and my brother, who have supported me emotionally and financially—is proud of me. I am blessed to have their unstinting support and never could have done this without their help. I thank each and every one of them.

I'm regularly asked how all this has affected me. I am much calmer and more patient, and I pick my battles, focussing only on things that matter. People say I've become more balanced. If I feel down or low, I now have a built-in mechanism to come out of it. I get up, dust myself off, and walk on. When my son could no longer walk, one of the most difficult things for us was his social isolation. Being unable to run and play with his friends meant he lost some of them. We thought of moving from where we lived, but he would have lost the few friends he had. And we too were getting socially isolated because our son was so dependent on us. Many of our old friends had stopped visiting us because we were caught up in our new life. But all in all, this experience has made me a better person.

People sometimes see me as an 'expert' or a social activist since I have informed myself so thoroughly about stem cell research and practice, even more than normal bio-medical doctors. The fact is that I can offer my son more than what all these so-called researchers can. I began doing this only for the love of my son, and if others benefitted, then so be it. I don't even want to go to any conference or academic gathering. In India, the reality is that nobody cares about a disabled child, and even if I were to narrate my story, nobody cares. It is ironic that India is a spiritual country but it has so much apathy. For instance, in Geneva or the UAE, some of the countries we have visited, people stop and help you in malls, in lifts. They push the chair and wait for you and only enter if there is enough space. Here everybody rushes in. I repeat—nobody cares, not the system nor the politicians. Living with a disabled member of a family is just one big obstacle in this country.

But for us parents, the life of our child is precious and it's a matter of life and death. We cannot afford to wait and do nothing. Our son had the fourth infusion of stem cells with hypoxic temperature cells, in Gurgaon. Although we don't see any improvements in my son, he is stable. Again, we are trying to delay the degeneration process. No one wants to see their son die. They say that what you resist persists. So my philosophy is to go with the flow and do what I have to do. Perhaps this is my own special spiritual journey, a trial by fire as it were. But we are living our lives and living them happily so far in spite of the roller coaster of emotions constantly being experienced.

We continue our endeavour with a hope in our hearts that at least we have done the best we can do. I wish parents like us could turn their grief into action by whatever little we all can contribute. I am sure we all can look towards a better and a healthier life and future. I am continuing with the therapy with a hope of delaying as much as we can and hoping for that eureka moment that brings a wave of health for all little angels to live as healthfully and normally as possible. The science of stem cells is very interesting; when we get hold of the specifics and learn to control this science, it will be a magic bullet for almost all medical treatments. The whole of the human body is made up of stem cells; hence, one can imagine the huge outcome of this science, which will be revolutionary in nature.

Ripudaman Singh has been working in the banking and financial field for more than 22 years. He holds a bachelor's degree in Commerce and a master's degree in Personnel Management from Pune University, India. He has had a very successful professional work experience in various leadership roles. He became involved with stem cells when his son, presently 16 years old, was diagnosed with Duchenne Muscular Dystrophy in 2005. He has been following stem cell research very closely for more than nine years and participates in various other research activities. Over the years, he has become an advocate for stem cell research and its huge potential, if practiced in the right way. He is passionately committed to finding stem cell–based therapeutic solutions for incurable conditions in children of all ages.

Accidental Events: Regenerative Medicine, Quadriplegia, and Life's Journey

Lola Davis and Shannon Davis

Regenerative medicine has been on the horizon for several decades with an acute awareness of stem cell and gene therapy approaches' potential to transform modern medicine. This field is not without controversy, ranging from objection to the very use of stem cells to means and ways of researching potential treatments. Nonetheless, pioneering researchers have taken many different paths in their approaches to examine and explore potential treatments/application of stem cell treatments for various conditions. Still, these treatments/applications remain remote. We read of clinical trials, but to date most of the research remains in the clinical stages.

The potential of stem cells to transform medicine will be a reality one day, but for families needing help today or yesterday, the immediacy of

L. Davis (✉)
Department of Education Emeritus, University of Oklahoma,
Norman, OK, USA

S. Davis
Dallas, TX, USA

© The Author(s) 2018
A. Bharadwaj (ed.), *Global Perspectives on Stem Cell Technologies*,
https://doi.org/10.1007/978-3-319-63787-7_8

needing to make decisions play a critical role. Parents of desperately ill or injured children, especially those for whom no established treatment exists, search for and are often willing to engage in experimental treatments with potential positive outcomes. When there is no hope, a glimmer can become a beacon in the tunnel.

Shannon

Scared, the car is rolling, grabbing the seatbelt from around my neck. Darkness.

'Hold on, honey, someone is coming to cut you out. They're almost here. Stick with me'.

'She is bleeding from her head right here'.

'Sweetie, we have your phone, who should we call?'

'The flight is on its way to take you to Baylor Hospital'.

Awake. Intensive care. Screaming loudly in my head that the stupid doctors don't know. Don't know.

On 25 August 2007, the most agonizing, heartbreaking day of my life, I became a quadriplegic. A multicar accident and rollover resulted in massive trauma to my spinal cord, causing it to stretch and tear from C-5 to T-8. The severity of the damage left me paralyzed from the armpits down. I had very weak, limited use of my right arm and no control of my left. The damage was so severe that I actually died as my vital systems failed at the accident scene, and I had to be resuscitated.

The shock, disbelief, and despair of awakening to such news were devastating. This deep despair has the potential to rob one of all will to move forward. I did not allow myself to think there was no hope. Nonetheless, I was in Baylor Hospital, Dallas, Texas, attached to medical machines keeping me alive and stable for ten days. Within three days of being in intensive care, I underwent a 12-hour spinal surgery to stabilize my injuries as best as the surgeon could. Following the surgery, a drainage tube was left installed to control the fluid buildup within my spine. I was afraid as I remained there for days, hooked to medical lines and a respiratory machine that breathed for me. Ultimately, I was discharged and transferred from intensive care to the rehabilitative ward of the same hospital.

The surgeon and rehabilitative doctors said I had no hope for recovery. They said I would never walk again. They said my spinal cord was stretched, torn, and dead in many areas. Their focus was on survival and then physical therapy to strengthen and improve the movement I still had. They said there were no medical treatments here or abroad. No hope to help me walk again. The best outcome they believed was that I could regain use of my arms and be able to conduct minimal feeding tasks.

However, I have been a fighter my entire life. I arrived in this world two months early, was discharged from the hospital at three pounds 12 ounces, and have always thrived by being challenged and pushing boundaries. I am independent, goal oriented, and persistent. A challenge has energized me throughout my life. These qualities reemerged and supported me as I then faced the greatest challenge of my life. In rehabilitation, I worked hard and struggled to gain strength and arm movements. My focus to improve each day sustained me, but I didn't accept that I wouldn't walk again or get better results than just the use of my arms.

My parents and I researched the web. We discovered that stem cell treatments were actually being conducted in various countries around the world. However, the doctors told us that all stem cell treatments were quackery and we were warned about dangers of stem cell outcomes such as teratomas or cancer and even the fraudulent use of unknown substances being injected. So, fearfully, we backed off from this idea.

Lola

At the time of Shannon's injury, we were but vaguely aware of stem cell treatments. Sporadic news accounts were published, but we had not paid much attention. Much of what was published at the time of her accident was not encouraging. Traditional physicians in the USA were very suspicious and against stem cell treatments, and Shannon's physicians repeatedly told us there was no hope that she would regain much after the first 12 months through any treatment program. One physician told her stem cell treatments were fraud and no one was conducting authentic treatments any place in the world. Like many parents facing such a devastating injury, we did not absolutely accept that perspective and conducted our own research, but we were cautious. We extensively researched online to find any help for someone with a spinal cord injury. Our initial research efforts were not encouraging: in fact, the results were dismal. We turned to other areas for help at that time.

Shannon's father, as a former coach, believed strongly in 'muscle memory' and the importance of exercise, especially repetitive exercise. With his coaching background, he believed that 100 repetitions of an activity instead of 50 would be an ideal practice program. Therefore, any new skill certainly had to be practiced regularly and repetitively. With such a perspective, it was no surprise that his research and strong focus led us to a special rehabilitation program entitled Project Walk, in California, USA. Project Walk was particularly interesting to Shannon and her father due to the emphasis on an extensive exercise regimen. The program also targets intensive assistive exercises to aid in recovery, increased mobility, and movement in spinal cord-impaired clients. Five months after the spinal cord injury, in January 2008, Shannon and I traveled to California to allow her to participate in Project Walk for a month. It proved to be a valuable, positive, and surprisingly informative trip.

Project Walk proved to be an exemplary assisted-exercise program, and Shannon gained strength and endurance during our month there. But the absolute highlight of the trip was meeting a young man who had been to India for stem cell treatments. This was incredible, amazing, and unbelievable news to us. After being told there were no stem cell treatment programs in the world, we were talking to a family who had attended and received stem cell treatments! 'M' had a spinal cord injury similar to Shannon's, and he and his father were excited to share their experiences in India and the positive outcomes. It was almost unreal that we were receiving the exact information we had spent months trying to acquire. We immediately began to research the information and found patient blogs, published information, and testimonials.

We are forever grateful to these new friends for introducing us to the work of Dr. Geeta Shroff, of Nutech Mediworld in New Delhi, as she has been engaged in stem cell treatments since 2000. Our initial research had not revealed her clinic, and we subsequently discovered that much of the information on the internet was firsthand because the clinic does not advertise or market the clinic and treatments. Nonetheless, we had a well-documented lead to an innovative stem cell treatment facility.

Shannon

In January 2008, after my release from Baylor Institute for Rehabilitation, my mother and I went to Project Walk for a month. Indeed, it was a special place. The therapists who worked with clients were athletic trainers with a focus on activating muscles and movements that were not working! Their goals and strategies were aimed at triggering nonresponsive movements and increasing mobility. The training was rewarding, and I gained some recovery. However, the ultimate highlight of our journey to California was meeting a fellow quadriplegic, 'M', who had gone to India for stem cell treatments. Specifically, he had received embryonic stem cell treatments from Dr. Geeta Shroff at Nutech Mediworld in New Delhi, India! Serendipitously, my time at Project Walk was destined to lead me to embryonic stem cell treatments. Five months after my car accident, I was face to face with a young man who had not only survived stem cell treatments but who also shared details of his improvement following treatments. 'M' was upbeat, positive, and encouraging about his experiences and treatments. Could I dare to hope?

Lola

As 2008 continued to pass, Shannon again started researching Dr. Shroff's work and outcomes and contacted as many former or current patients as she could to discuss their treatments and outcomes. She was hopeful from speaking to others but was fearful of the warnings from traditional physicians. Toward the end of the year, I told Shannon 'enough'. She had found all the published information and spoken to numerous patients, and it was time to make a decision and take action.

Shannon

During the remainder of 2008, I continued to work out, acclimated to venturing from the house, returned to work part time, learned to overcome obstacles, and tried to recover my health. I also became serious about researching stem cell treatment options. At that time, most of my online searches did not result in positive information about stem cell treatments. The scientific literature was replete with dire warnings of possible negative outcomes, and the major concern was that stem cells would cause cancer or develop into a cystic teratoma. There were warnings of unethical stem cell treatment facilities in unregulated areas of the world, and warnings abounded that clinics were in business just for profit without regard to patient safety. The research was not encouraging, and my parents and I resolved to attack my injuries throughout the remainder of 2008 with exercise and the promise of improvement from Project Walk.

But I could not stop thinking about stem cell treatments. The human-science aspect of stem cells just seemed the most logical route to help heal my spine. I refocused and decided to get stem cell treatments. I once again conducted extensive reviews of current treatment protocols in published literature or online venues and found five countries offering treatments. I read, contacted the clinics, and attempted to locate patients who had received treatments. I had few concrete responses from patients of most of the clinics but was able to meet with a young man who had gone to Germany for treatment. He had some small, positive results but nothing to help him regain motor function. He did not plan to return to Germany for follow-up treatments.

My research quickly became refocused on Dr. Geeta Shroff, New Delhi, India. Not only had I actually met a current patient of hers, but also I had received numerous responses from other patients of hers. It became readily apparent that the only authentic treatment option in the entire world was Dr. Shroff in India, and I continued to extensively review the numerous postings of her past and current patients. She had been utilizing her protocol for almost a decade, and positive results were well established.

But still, I was afraid. Going to an unknown clinic and foreign country and as a quadriplegic was a daunting task. But as a family we decided it was my best chance. In early 2009, I contacted Dr. Shroff to discuss my case and seek admittance to her program. She was amazingly supportive and thorough in her response, so I submitted the required medical records and anxiously waited for a response. Positive news soon arrived: I was accepted as a patient.

Lola

We had engaged in an extensive review of Dr. Shroff's work and concluded it was at the cutting edge of the regenerative-medicine realm. We
recognized her as a medical pioneer, an impassioned visionary with a
proven track record of improving desperate lives. We were encouraged
and willing to participate because we found her methodical and documented approach consistent with clinical case studies. The methodology,
medical facilities, and staff were professional. Most important, Shannon
had communicated with many of Dr. Shroff's patients before engaging in
treatments and was fully informed of the potential outcomes and expectations. Dr. Shroff kept Shannon and us informed and updated on all
aspects of the spinal cord treatment approach. This honest exchange of
information and the professional safeguards in place were a foundation of
our decision and commitment to travel to India and seek treatment. We
determined this was our only hope in the entire world. That was a daunting thought—the entire world—but we believed and placed our precious
daughter in Dr. Shroff's care.

The next phase of our lives involved fundraising for the full three
months of initial treatment. This was a bit daunting in 2008 and 2009 as
well: many people were and are against stem cell treatments. We shared
our story with a few select friends and family, and soon, very soon, we
had an energetic team that set about forming fundraisers to support trip.

Journey to India

We arrived in India at the end of January 2010 and stayed until mid-
April. We traveled with six checked bags, six carry-on bags, and Shannon's
wheelchair—no small task. After a 24-hour flight, we were feeling quite
alone and exhausted, but as soon as we emerged from the airport baggage
area, a staff member from Dr. Shroff's clinic was holding a placard with
Shannon's name. We were safe and ready for our next phase.

Our next phase was the shock of how small Indian taxis are, but our
driver managed to store all our bags inside and on top of the cab. As the
first few days unfolded and we went to various medical clinics for essential MRIs, X-rays, and so on, we experienced the ingenuity of Indian

people to solve all and any dilemmas a girl in a wheelchair might encounter. This was my first trip to India, and I encountered many shocks as we became immersed in amazing, incredible India. The beautiful, colorful clothing, rich various languages, spicy food, unbelievable traffic, and open markets were so very different from our home. We were in awe of the country, but the offers of help from everyone we encountered made us feel not quite so far from home.

When we first arrived at the clinic, in the wee hours of the morning, Dr. Shroff's professional team took over. Shannon was immediately seen by the nurses, and a doctor came after breakfast. Dr. Shroff greeted Shannon that morning with the medical team responsible for Shannon's treatments, and my fears completely disappeared. I knew we were safe, in a professional medical clinic, and Shannon was receiving the most advanced stem cell treatments in the world. All our care, food, and lodgings were included in the treatment cost so that we didn't have to worry about daily events. Dr. Shroff used her professional medical knowledge of more than a decade to design stem cell protocols for Shannon and others under her treatment. Shannon received stem cells via muscle, infusion, and spinal cord as well as topically when needed. In addition, she had two sessions of physical therapy and one occupational session daily. This approach was holistic, with concerted care given to all aspects of her recovery.

Immediately after her first spinal injection, Shannon had new sensations in her toes, legs, and back. Words do not and cannot adequately express the joy, reverence, and deep gratitude when the hope of healing becomes a reality. Hope is hard to face when there is no treatment, but with the first spinal injection and Shannon's response, hope was transformed to reality.

Shannon

I was given the best care in a clinic with full-time medical staff on site 24/7. The treatments began the very first full day I was there, and the full protocol was explained in depth, preparing me for what was to come. In addition to embryonic stem cell injections, I was engaged in daily specific, targeted physical-training sessions directed by doctoral physical therapists, with the exception of the days I received spinal injections and on Sundays.

Lola

When Shannon arrived for her first treatments, she was still using a seat belt to keep from falling out of her wheelchair, and she was supported by armrests and side panels (to keep her hips aligned). As a quadriplegic, she was paralyzed from the chest line down and could not remain upright without using armrests. She was in a pushchair when we arrived, but I had to push her most of the time because she didn't have the strength or endurance. Also, standing would cause her blood pressure to crash and she would faint. During the three months of treatment, Shannon gained enough core body strength to be able to discard the seat belt! Her back muscles had already been stronger, but her abdominal muscles began triggering as well. This was and is a major milestone for a quad, as many are not able to get out of a power chair into a pushchair. She was also using back muscles to aid in upright support, her muscle tone was increasing in all the affected areas, and her arm strength was increasing. She was showing improvement in all muscle groups. She was also able to stand upright with leg and abdominal calipers for longer and longer periods. Even before we left in April, we were planning a second trip for later in the year. She was experiencing so many positive results that we couldn't leave without making our plans to return.

We returned in September 2010 and have gone back at least once per year, since with the exception of 2016. Shannon has made nonstop progress since the inception of treatments. She continues to gain and will remain a faithful patient until she is walking. We are told that stem cells go to injury sites and have the potential to last up to five years. She has the same sensations she had on the first trip and has added many more. Although they vary in intensity, they remain in all areas from top to bottom. Her core body strength allows her to engage in physical training designed for paraplegics rather than quadriplegics. She started upright physical therapy in a full body brace and calipers and can stand without a body brace and move her legs. She also has balance, can use weights with upper arms and core, and engages in facilitated walking.

We all have life journeys that are impacted by others. Sometimes we are very much aware of the influence that others have, sometimes not. Our family is acutely aware, and we are beyond fortunate to have had

Dr. Shroff as a primary healer for our family and daughter. We have witnessed and experienced the healing power of her pioneering work and stem cell protocols.

Today, Shannon works three-quarters time. She dresses herself, does her hair and makeup, gets into her adapted van, transfers to the driver's seat, drives to work, drives to physical training, drives to wherever she wants to go, and drives home at the end of the day. Her life has become as normal as it can. She still uses a wheelchair, but her total body recovery has been phenomenal, and walking solo remains her final goal. She doesn't give up.

Shannon

I have never looked back and know the decision to engage in stem cell treatments has transformed my life. My regained muscle recovery, mobility, endurance, and overall health are a direct result of the stem cell treatments. I work for 75 percent, drive, dress myself, and go wherever I wish. My independence is a blessing.

The decision to participate in embryonic stem cell treatment was based on extensive research. My family and I came to the realization that Dr. Shroff, her clinic, and overall protocol were unique in the world at that time. The rigorous medical attention to treatment protocol was a deciding factor for me. Even today, I am planning my next trip to India, as I believe there is no other place in the world to help me.

Lola Davis, Ed. D., is Professor Emeritus in Oklahoma, USA. She was a full professor and member of the Department of Curriculum and Instruction from 1992 to 2008. During that tenure she served as department chair, curriculum coordinator, and content lecturer; authored and/or participated in four national accreditation reviews for the early childhood program; was a member of university level policy and guidance committees; and chaired a subcommittee at the university level on University of Central Oklahoma's steering committee to attain national level university accreditation from North Central Accreditation. Previous to her time at UCO, she taught in a number of Oklahoma public schools and served as an administrator for an elementary school. Davis' doctorate is in

Curriculum Instruction Elementary Education with a specialization in Early Childhood Education attained in 1993 from Oklahoma State University, Stillwater, Oklahoma. She lectured the undergraduate courses: Language Development and Emergent Literacy, Psychological Aspects of Multicultural Education, and Creativity. Graduate courses focused on action research methods in constructivist practices of early childhood teachers and curriculum theory for young children aged 4–8. Her research interests include teacher development, action research methodology, constructivism (epistemology of how young children construct knowledge), and very young children's emergent literacy and logicomathematical development. While serving on the faculty at UCO, Davis' publications focused on the essential role of professional reflection in teaching practices and perspectives, challenges and opportunities for teacher education departments, and the constructivist foundation of children's cognitive development in literacy and logic. She was invited to lecture, key speak, and train professionals at local, state, national, and international levels including lecturing and participating at an international Oxford Round Table on Emergent Literacy, Oxford University, UK, 2005.

Shannon Davis is a credentialed CPA (Chartered Accountant) and PMP (Project Management Professional). Her specialization and work focus includes forensic and fraud accounting, external auditing, internal auditing, Sarbanes-Oxley, and business process. Her consulting experience includes compliance adherence; business process efficiency improvement; business process reorganization; and fraud prevention/control implementation. She is a global training facilitator for interns, new auditors, senior auditors, and new managers for two Fortune 500 companies. She has served as member of a Global curriculum development task force for interns-partners for an international Fortune 500 consulting firm. Her industry client focus has been on Fortune 500 Security and Exchange Commission manufacturing clients in $1–16 billion range. She has been in the profession for 21 years and is an acting global audit director for four publicly traded companies.

Biocrossing Heterotopia: Revisiting Contemporary Stem Cell Research and Therapy in India

Nayantara Sheoran Appleton and Aditya Bharadwaj

Introduction

As I looked on, Dr. Bhatia and his attendant opened the Styrofoam box full of liquid nitrogen. As the vapors rushed to escape the confines of the box they had travelled in for the past four hours from Hyderabad to Delhi, Dr. Bhatia reached his gloved hands into the box and pulled out two vials. They both (the doctor and his attendant) then took each vial and gently started rolling them in their hands to thaw out the millions of frozen cells that had travelled across the country for a patient that was in the next room. Dr. Bhatia, while rolling the vial in his hands, continued to tell me about his patient, a lower-middle-class woman in her mid-fifties with optic atrophy who had tried all treatments and had recently turned to stem cell therapy as her last resort.

N.S. Appleton (✉)
Cultural Anthropology, School of Social and Cultural Studies, Victoria University of Wellington, Wellington, New Zealand

A. Bharadwaj
Department of Anthropology and Sociology, Graduate Institute of International and Development Studies, Geneva, Switzerland

© The Author(s) 2018
A. Bharadwaj (ed.), *Global Perspectives on Stem Cell Technologies*,
https://doi.org/10.1007/978-3-319-63787-7_9

195

The doctor took the thawed cells and moved to the next room. Mrs. Padma lay on the single bed while her husband waited outside. She seemed visibly distraught about the insertion she was about to undergo, and as Dr. Bhatia assured her, he asked me to stand above her headrest, as I would get a 'better view' of the insertion. Instinctually, Mrs. Padma reached out and caught my hand; I decided to move closer and offer the support she needed as I watched Dr. Bhatia transfer the cells from the vial into a syringe with a specialized two-inch needle, used for optic nerves. During the transfer and organization for the insertion, Dr. Bhatia, while focusing on the process of what he was doing, continued to talk to me. (It almost seemed he did that to assure Mrs. Padma and perhaps me, about the everydayness of the procedure for him). As he took the needle and carefully inserted it into the patient's eye, he explained about stem cell therapies: 'This is what is being called a 'cocktail' of stem cells. These are not pure bone marrow-extracted cells that have been centrifuged, but rather cell lines that have been developed particularly for optic nerves. These aren't your embryonic cells nor the autologous cells that everybody now chooses to work with but somewhere in between because I find them more effective'. (Appleton field notes, 2014)

This conversation happened in early 2014. Since then, there have been numerous changes in clinical practices around therapeutic application because of the shifts in regulatory frameworks in India. The key factors driving changes in the field are the 'guidelines' offered by the Indian Council of Medical Research (ICMR) and Drug Comptroller General of India (DCGI) about permissible stem cell research and therapies (ICMR 2007, 2013). One of the direct implications of these regulations has been the shift in focus of both the research labs and clinical facilities to move away from human embryonic stem cells (hESCs) toward somatic stem cells. As Dr. Bhatia explained, 'We both know that Mrs. Padma's chances for her sight to return would be higher if we used embryonic stem cells, but I don't want to take on the government and the entire research world with what they have decided is outside the realm of acceptable stem cell therapies'.

He explained that he did not work with embryonic cells, because of the regulatory and scientific world viewing them so critically, but he did not like working with autologous stem cells because he found that efficacy was limited and at times minimal. He felt that most clinics and physicians were happy to work with minimally manipulated autologous cells, since they were 'non-risky' and provided a safe way to 'new medicine'.

Over the 2 years that Appleton has worked with Dr. Bhatia, he has continued to work with allogeneic somatic cells he purchases from a lab in Hyderabad. However, as the regulatory framework pushes for the guidelines to become law in the next year or two, he suggests he will have to develop his own lab where he can seek clinical-trial funding to continue to do the work he does for his patients.

As research that drives stem cell therapy crosses the terrain of hESCs and moves toward cells derived and developed from somatic cells, it is perhaps timely to examine, following Sarah Franklin, the contours of stem cells as they are normalized and made 'curiousiour and curiousiour' (2013). The emerging politics and science behind the curious shift from embryonic to somatic cell research in India and the push to mainstream autologous somatic cell transfers therapies is a good example of 'biocrossing' (Bharadwaj 2008). That is, transfers achieved through twin processes of extraction and insertion and administered as an intended medical resolution of a pre-existing social or medial problem. Largely, biocrossing can be a conceptual or real movement between biology, biology, and machine and across geopolitical, commercial, ethical, and moral borders of varying scale (Bharadwaj 2008, p. 102). The notion of 'bio' implicit in this movement or 'crossing' is doubly articulate. First, bio is quite literally a biogenetic substance saturated with political, ethical, therapeutic, and commercial value accessed through these twin processes. Second, the notion of bio signals the presence of an implicit and explicit individual and/or institutional biography inextricably (re)written as crossings gain momentum. In this chapter we articulate the faint traces of utopic and dystopic logics underscoring these 'crossings' and the evolving biography of a contested terrain this (re)scripts. In so doing we engage with our ethnographic immersion into the lives of physicians, researchers, policymakers, and patients to conceptualize evolving scenarios that remain divergent and yet the source of emergent but shifting utopias and dystopias that get mirrored and experienced as a heterotopia.

Biocrossing and Heterotopia

Medical anthropologists and science and technology studies (STS) scholars have started looking at stem cell technologies and therapies as a way to understand and unpack the complexities of the social lives of this latest biomedical intervention, which as a nascent science has managed to mobilize capital and labor (both specialized and nonspecialized) in a geopolitical moral battle (Franklin 2006, 2007; Bharadwaj 2012; Thompson 2013). Thompson's rich ethnographic account, for example, focuses on a time period she dubs 'the end of the beginning of stem cell research' (2013). What she refers to as the 'end of the beginning' of stem cell research coincides with a shift from ethical issues surrounding hESCs to a stem cell science based on somatic (adult) cell lines and autologous cells. Thompson's cautionary note about ethics in stem cell technologies is important when she writes, 'The end of the beginning of stem cell research must open up, not close down, what can be raised as ethically important in the field' (2013, p. 27). As we track the 'biocrossing' from embryonic to somatic cells, we must open up the conversation not only on ethics, but also on the curious way this crossing is enabled and its implication—both for science and society. We suggest that the undulating landscape of stem cell research and therapies in India is a curious mélange of utopian views of benign good science of cellular therapies offering 'cures' for some of the worst known intractable afflictions and dystopian fears of runaway bad science violently proliferating dangerous cellular interpolations. To a large extent the moves to facilitate a shift from embryonic to somatic cell research in India mirrors this curious mélange. The utopias (and dystopias) shaping this curious terrain can be understood as 'sites with no real place' (Foucault 1967, p. 24). According to Foucault, both utopias and dystopias are 'fundamentally unreal spaces' (p. 24). However, he does allow for 'real places'—discursive and concrete—in civilizational and societal contexts that are something like counter-sites, an effectively enacted utopia in which 'other real sites that can be found within the culture, are simultaneously represented, contested, and inverted. Places of this kind are outside of all places, even though it may be possible to indicate their location in reality'.

Foucault describes these places/sites as absolutely different from all the sites that they reflect and speaks about and christens them as heterotopias, the contrasting other of utopias. For Foucault these places contrast utopias significantly and are absolutely different from all the sites they reflect and speak about. In this formulation a contrasting figure to utopia is not dystopia, but rather a reflection of the utopia itself. It is seemingly real, connected to the utopian ideal and/or projection and yet unreal as it can only be perceived as mere approximation, a reflection of the utopic and everything material and otherwise that surrounds it. A heterotopia is a real, existing 'other' place that can be experienced. They are counter-sites within a culture, enabling life to carry on functioning in a non-normative vein in the face of normative circumstances.

The most important question is what a heterotopia reflects. We suggest that a heterotopia is perspectival. It can conjure and seemingly concretize in space- and time-enacted utopias and dystopias. In other words, a heterotopia collapses the distinction between a utopia and dystopia to the extent that the reflected real is prone to mutate based on the concrete reality of the reflected site. The reflected real momentarily stabilizes to birth a perspectival reality. In other words, mythic and real as well as utopia and dystopia collapse and stabilize to form perspectival realities. Kevin Hetherington (1997) sees heterotopias as spaces of social ordering that are different. These spaces, he argues, can be transgressive or hegemonic. In the end, heterotopias are made up of multiple and often 'incongruous processes of social ordering' (Street and Coleman 2012, p. 9).

Stem cells can be reimagined as heterotopias: manifest entities and discursive sites suffused with real and imagined, and utopic and dystopic alterations made manifest as biocrossings gain traction between the biogenetic, technoscientific, socioeconomic, and geopolitical landscapes of possibilities. Like a mirror image of the seemingly real, these cellular heterotopias are spaces that seem hegemonic but in practice are condemned to operate in a nonhegemonic, inconsistent manner. In this respect, biocrossing a heterotopia produces concrete social spaces fraught with opportunity and danger that on occasion can be calculated risk or a forced dislocation as the last resort (Bharadwaj 2008, p. 111–112). The biocrossings undertaken by actors in India are indeed complex moments that allow for a nuanced analysis, as they are not singular occurrences that happen symbiotically, automatically, or 'naturally'. These biological and local biographies depart from the purportedly real, and it is this slippage that needs more focused analytical work.

Ethnography

This chapter is based on research conducted in India from October 2013 through December 2015. The data presented here is, in many ways, preliminary and a precursor to some of the realities and arguments that may emerge as we continue with this work in the future. It includes participation observations and informal interviews Appleton conducted with interlocutors in cities in India: Delhi, Mumbai, Bengaluru, Pune, Hyderabad, and Apela. While Delhi and Mumbai are 'tier-one' cities, Pune, Bengaluru, and Hyderabad are 'tier two', with all boasting multiple 'stem cell clinics' (irrespective of whether they are doing lab research or patient therapies). The last of these cities, Apela (a pseudonym, since it is a small town with easily identifiable clinics where Appleton worked), is in the western part of the country with a small hub of clinics and faculty at a teaching hospital involved in stem cell research and therapeutics. Appleton did most of her clinical participant observations in two hospitals in Mumbai that specialize in stem cell therapies; with two physicians in Delhi who worked out of different hospital operation rooms; one leading hospital with a top-of-the-line research lab and facility in Delhi; one clinic and one lab in Apela; shorter visits to two clinics in Pune; and multiple physicians, clinicians, and researchers in all these cities. In the course of the research, she spoke to over 100 participants (some of them multiple times) and spent 2 years fully immersing in the everyday lives of patients, physicians, clinicians, and policymakers involved with stem cell research and therapies in India. Bharadwaj's research has mainly focused on the emergence and spread of stem technologies across India. His research, supported by the European Research Council, engages with the scientific, policy, and everyday experiences in culture and therapeutic nurture of stem cells attracting global traffic in patients suffering from a range of incurable and terminal conditions to India.

The data in this chapter is drawn from a larger research project supported by a European Research Council grant (#313769). In this chapter, all names are anonymous to preserve confidentiality, per the ethical protocols at the Graduate Institute of International and Development Studies, Geneva, and the European FP7 framework guidelines. All respondents were informed about the nature of the research project and their ability to withdraw at any point from the study. Further, all research ethics protocols per the European FP7 were rigorously followed.

Biocrossings and Regulatory Frameworks: Physicians

While there are more registers to examine when studying the biocrossing(s) from embryonic to somatic, we focus on regulatory frameworks in this chapter as a way to examine the role of one scientific artifact from various perspectives. Talking to patients, physicians, and policymakers, it is evident that the stem cell terrain in India is indeed very complex, with multiple stakeholders (with new complexities and stakeholders emerging every day), so the focus here on regulatory frameworks is just the start of a conversation rather than an attempt to foreclosing boundaries. In this section, we look at the role of the state in promoting this latest of biocrossing, by privileging one form or therapy over other. The current regulatory 'guidelines' in India—while providing various ways physicians, researchers, and clinicians could develop and use stem cell therapies—had clearly marginalized hESCs as 'unethical', 'non-permissible therapies', and 'dangerous'. The DCGI and ICMR made certain forms of cellular permissible but are not willing to even remotely regulate but rather outright make hESCs outside their purview set the tone for how the country discusses stem cell therapy. An automatic 'good/permissible' science versus a 'bad/rogue' science has been established. While this establishes a certain utopian and dystopian hierarchy, heterotopic topography, these moves produce, often as an unwitting corollary, destabilizing effects.

Let us examine the responses to two documents written in 2013 by the Ministry of Health and Family Welfare (MoHFW). The first is a draft guideline issued in February 2014 by the Central Drugs Standard Control Organization (CDSCO) of the MoHFW, Government of India, called 'Guidance Document for Regulatory Approvals of Stem Cell and Cell Based Products (SCCPs)' (Guidance Document henceforth) (Central Drugs Standard Control Organization 2013). The other was issued by the ICMR and is the 'National Guideline for Stem Cell Research' (2013). Since they were both relatively new, often the people Appleton spoke to conflated the information in these documents.

One of the key issues of concern was that in the latest version of the second of the documents ('guidelines' in the remainder of this chapter), the authors and policymakers had removed the word 'therapy' from the title. The 2007 'guidelines' issued by the ICMR had 'Therapy and Research' in their title as a way to provide guidelines for researchers and clinicians involved with therapeutics along with research. However, according to the 2013 'guidelines', the term 'therapy' had been removed to the effect that anyone conducting therapeutic stem cell work was effectively involved in malpractice. Rather, the other document, issued by the CDSCO ('guidance document' in the remainder of this chapter) became the guiding point for physicians involved in clinical therapeutics with stem cells. If you were a physician working in any capacity to provide stem cell therapies, you were no longer under the purview of the 'guidelines'. Between both documents, these physicians and their work were now under the governance of the DCGI office, effectively labeling their stem cells as 'drugs' that needed to comply with the Drug and Magic Remedies Act of 1954. This limited their abilities to conduct 'cutting-edge research', since any 'drug' had to go through several phases of very expensive clinical trials before being approved.

The guidance document was issued just before Appleton attended a conference on stem cell therapies in Mumbai, and the tension was palpable at the conference as various clinicians tried to figure out which side of the law they operated on (even though these were not legislations but rather 'helpful guides for ethical' stem cell development in India). In large part these interlocutors simultaneously appreciated and bemoaned these documents as a foretelling of what was to be the future of stem cells in India. However, the seeming hegemonic oversight crumbles when its enforcement is scrutinized. None of the above guidelines and guidance documents can be legally enforced. The negotiation with the state and its organs such as ICMR and DCGI remains contingent and context sensitive with tremendous elbowroom for individual and collective bargaining, petitioning, and expedient subversion.

While hESC research continues on a global level, the current regulatory and state-funding environment has made such research (and therapies) a fringe endeavor in India. Often the response would be to point out that in the post-Bush era the funding for hESCs research has been permitted. But in India, it is still considered too volatile to touch. One physician joked, 'You think their embryos are better or less volatile than ours'. He went on to explain, in detail, how the lack of funding for embryonic and fetal research leads to lack of true innovative work in India. He pointed out that physicians and researchers wanting to work with embryonic and fetal cells had crossed over to working with 'simplistic' autologous bone marrow transplants as a way to stay in the 'business' and support their practices. Given that the Indian medical establishment is largely privatized, physicians and clinicians pay their bills by performing these particular therapies and publish these results in academic and scientific journals, which in turn means they become specialists in those treatments versus being able to take on more innovative research. Yet, because heterotopias are inherently plastic and adapt at bringing together several incompatible sites, hESCs in India, as chapters in this volume amply testify, are truly thriving and producing dramatic results.

Biocrossings and Regulatory Frameworks: Policymakers

The other side of the debate about crossing over from embryonic to somatic autologous cells was composed of the policymakers working toward situating Indian stem cell research and therapeutics on an international platform of respectability and recognition. This was a goal quite similar to those of the physicians, who also wanted India to be the forerunner in this nascent medical innovation. Of course, although the end points were the same, the policymakers' relationship with the physicians was quite a contentious one. The two main problems identified by the policymakers in regard to stem cell therapies were that some physicians were providing 'unproved' medical treatments at very high costs and no safeguards were in place for patients who might not benefit or, worse still, suffer from negative consequences of these experimental therapies.

When talking to policymakers about why embryonic stem cell research and/or therapies in India were not being recognized (and thus perhaps regulated), one of the former members of the regulatory bodies pointed out that what the ICMR and the Ministry were doing was for the benefit of the science itself. She gently reminded Appleton in the interview,

> See, nobody understands this, but every regulation that is put in place is not to restrict science but to protect and enhance it. When these policies are put in place, it is not to punish 'bad' medical practitioners but to prevent the 'good' ones from getting a bad name because of the others. If not controlled now, and if India gets a bad reputation for providing dangerous treatments, then nobody … not one single doctor will benefit. We are trying to protect the field of stem cells by putting regulations in place and using international ethics as our guiding principles. We want India to be a place for the best medical treatments, both for Indian and non-Indian patients.

For her, safeguards against hucksters of stem cell therapies prevented the entire Indian medical community and medical tourism enterprise from suffering in the future. Again, a risks and benefits analysis formed the framework, where the risks needed to be minimized in the short term to ensure long-term benefits. hESCs proved to be riskier, and crossing over to autologous cells was one way the state minimized/mitigated its risks while being able to participate in the benefits of being an aspirational 'scientific hub'.

This conversation was held alongside other conversations about protecting the financial wellbeing of 'poor' patients who were desperate for a cure, but at no point was the issue raised of providing this form of personalized medicine at government hospitals or government-subsidized prices. It should be noted that, historically, the health budget in the GDP has been shrinking, and in the 2015 budget it was reduced to the smallest slice of 1.2 percent of the GDP (Rajagopal and Mohan 2015). The reality of India's public-health sector constantly shrinking and becoming ever-dependent on private health providers, international aid, and philanthropic agencies (each with their own problematic agendas) has implications for stem cell therapies. Even though some preliminary work (following some of the most stringent international standards of ethics and medical development) was ongoing in government institutions in India, policymakers' focus was on private stem cell clinics, hospitals, and institutions. The focus remained on 'enhancing' these spaces by encouraging them to operate within internationally established norms rather than focusing on enabling the government-sponsored stem cell to excel in order to provide personalized medicine to the largest portion of India. The 'poor' within this framework were available as docile experimental bodies but never viewed as worthy citizens deserving top-tier medical care from their government-medical establishment. The tensions were real. The aspirations of the medical community alongside the policymakers' were palpable. The biocrossing from one form of cellular therapy to the other was not an 'organic' move but a calculated risk the Indian state promoted/approved in order to mitigate future risks.

The state, in its endeavor not to be dubbed a 'rogue nation' and continue to make itself available as a site for scientific endeavors (in terms of attracting global capital for clinical trials, pharmaceutical intervention, etc.), regulated and disciplined itself along global logics of acceptable and permissible science. In conversations with clinicians and physicians, one would often hear grumblings about the US FDA's and US pharmaceutical companies' vested interests in not allowing hESC research to continue outside Euro-American labs, so as to maintain a monopoly on biomedical breakthroughs. However, when bringing up these issues with policymakers, the focus was often on safeguarding the poor and protecting the image of the nation while promoting India as a 'safe scientific space' for global science.

The inherent need to encourage one form of cellular research and therapy as safe and 'manageable' while deeming the other as 'dangerous', 'rogue', and 'unmanageable' makes visible the geopolitical machinations that drive this latest biocrossing. This self-disciplining and regulation allow for a crossing that eventually appears non-problematic and 'natural' while gradually erasing the tensions and debate that drive the science in the global biomedical market. It is not our intention to suggest that either embryonic or somatic cellular therapies are better or worse than the other or that one should be encouraged or discouraged, but rather to show how this latest biocrossing has been naturalized and left un-problematized by the state and following it, media, its publics, and even the local medical and scientific communities.

Further, we do not suggest that all physicians, clinicians, or policymakers thought similarly about crossing over from embryonic to somatic cells. Quite the contrary was evident in the research, as a majority of the physicians who worked with autologous somatic stem cells (i.e., one's own adult cells) thought their therapies were clearly superior and safer (largely considered superior *because* they were safer). Rather we focus on the contention above as a way to show the tensions that were impacting the naturalized crossing of one particular form of cellular therapy over others. What one group viewed as dystopia another articulated as utopia. The resulting heterotopia reflects these tensions that continue to author the biography of stem cell science in India.

Biocrossings and Regulatory Frameworks: Patients

Nowhere was the dichotomy between the dystopic futures bought on by cellular therapies versus the utopic potential of said therapies more pronounced then in the patient and patient-advocate narratives. The imagined utopic futures ranged from articulations of being able to gain access to stem cells from pharmacies, to being able to participate in everyday life by patients receiving or aspiring to receive stem cell treatments. On the other end of the spectrum were criticisms from patients and patient-advocate groups that found embryonic stem cell therapies 'experimental', risky, and without benefits. Often, narratives of patients whose therapies had not worked and had felt violated materially and beyond were reported in newspapers (Jayaraman 2014), which along with other critics imagined dystopic futures for patients receiving these therapies. These dystopic futures included fears of mass cancerous growths in patients who could not afford to treat and/or manage those future diseases. Interestingly, the dystopic futures often included future fears of different illnesses and current financial burdens on patients but not any concerns about the immediate negative effects of stem cell therapies. An enduring irony underscores these fears. The patients and advocates voicing them had not tried hESCs in India. In large part, the terrain of fear and anxiety was built up on purported evidence from globally dispersed sources of normative science that the Indian state in turn resurrected as proof for its regulatory concerns and a need for a calibrated shift from embryonic to somatic.

However, in between these extremes of people who either imagine stem cells as absolute cures or medically impossible 'scams', are hopeful and ambivalent patients often described as being duped into embracing stem cell therapies based on 'bad name science' (Bharadwaj 2015). As Sarah Franklin reminds us, all technological breakthroughs are imbued with certain levels of ambivalence. She writes,

> The ambivalence that characterizes the IVF encounter, while specific in its form to IVF treatment, is also more generic, and I refer to it throughout this book as 'technological ambivalence', arguing that it is a constitute component of biological relativity. As many social scientists have noted, such as Ulrich Beck (1992), ambivalence is one of the defining characteristic of the modern relationship to technology—be it television or email, robotics, or biotechnology, electric kettle, or plastic bags. (2013, p. 7–8)

The heterotopia of embryonic and somatic autologous cells reflects this form of ambivalence. The emerging regulatory attitude also reflects globally established ambivalence toward human embryonic source of cells and mythic fears of inherent dangers clinically interpolating such cellular entities. To cross this terrain is to both witness the emerging biography of the political anatomy of hESCs as well as geopolitical interests on the intersections of capital, science, and the state that favor one particular discursive production over another. True to form, the resulting heterotopia of stem cell remains both closed and open (Foucault's fifth principle of heterotopia), thus making the terrain both isolated and open to newer future permutations of biocrossings.

Popular narratives and global scientific discourse suggests that hESC therapies in India operate in 'unregulated' ethical and medical terrains (Sleeboom-Faulkner and Patra 2009), a charge provoking the search for a more disciplined form of cellular therapy, such as somatic autologous cell therapy. Thus, we can argue that all interested stakeholders involved in the production of a particular biotechnological innovation do not so much experience ambivalence but rather gradients of uncertainty. This in turn allows one particular lobbying or interest group to impact the technology and shape its heterotopic present and future. And it is in this heterotopic space the contestations are lively and important, as we know from previous scholarship on science and technology, that debates at these times of transition shape futures of technologies (Winner 1980). Patients and patient advocates, particularly those who sought out stem cell treatments in the absence of any other options for improving their conditions, were often not ambivalent about the somatic autologous cell therapies they undertook, but rather felt definitive about their decision to *choose* one form of therapy or clinic over another.

This was evident in many meetings with patients across the country. For example, Mr. and Mrs. Vishand Deb had come from Mumbai with their 13-year-old son (Sushant), who had been diagnosed with autism at the age of 6. They had chosen to work with autologous somatic cells rather than embryonic because of the 'less-risky' nature of somatic autologous cells, as they were their own son's cells coming back to him in an enhanced form. The state-supported discourse around the riskiness of hESCs had clearly taken hold and to a large extent had shaped the eventual treatment modality. The first round of stem cell therapy for Sushant was at the age of 9 that started 5 years earlier; he showed reduced signs of aggression and verbal outbursts, could be asked to do chores around the house, related to his parents, and often hugged and kissed his younger sister fondly. Over the 5 years of treatment, the parents became advocates for autologous stem cell therapy because of his improvements (while considering that his symptoms have not deteriorated as he grows older). They particularly felt comfortable advocating autologous cells after making the initial decision not to seek out hESC therapies.

With each decision by the Indian regulatory bodies, news stories reporting these regulations, the reconfiguration of physicians and clinicians to meet the regulatory frameworks requirements (i.e., not working with embryonic cells but rather autologous cells under certain conditions), the lack of availability of clinics performing non-autologous cell therapies, the increase in the clinics performing somatic autologous cells therapies, the reporting of these claims and efficacy of their therapies, the Deb family feel validated in their choice. They, like many of the patients and patient families Appleton spoke with, may or may not have been ambivalent (they don't remember) about embryonic over autologous cells therapies at the start of their 'search for cures', but grew to feel rather strongly about preferring autologous over somatic cell therapies. The idea that patients and patient advocates symbiotically chose one form of therapy over the other or are ambivalent in their decision-making process is not evident in our data. Rather, what is evident is that particular geopolitical motivations created a choreographed moment where particular forms of cellular therapies were deemed problematic. This had the intended effect of creating spaces for alternative forms of therapy thriving and creating patients and patient advocates for particular treatments. This in turn created publishable scientific data in forms of studies and number of patients being treated, creating public opinion (both global and local) that autologous stem cell therapies in India were under the purview of regulatory bodies but clearly safer than the dystopic futures promised by unregulated embryonic stem cell therapies. Yet, as we have seen in foregoing chapters, this sharp distinction is somewhat unsustainable. The emerging biocrossings reflect how the every distinction between embryonic and somatic has become a product of conscious policy and its discursive reverberations rather than being based on tangible data on hESCs lacking efficacy or being inordinately riskier than somatic cell transfers.

Within STS, looking at scientific knowledge/breakthroughs at moments when the debate is most intense about the future of that particular scientific 'discovery' allows us to see that the shape of scientific and technological 'progress' is not inevitable; it is a result of political decisions. Langdon Winner proposes that the moment of introduction of a particular technology is a moment of contemplation and debate about the eventual benefits of that technological innovation (Winner 1980). At the moment of introduction into wider markets, the politics and cultures that lead to scientific knowledge and technological innovations should be examined. For Winner, the technologies and the technological artifacts contained within them politic for two reasons: first, for settling within communities debates about what technology to adopt, and second, man-made technologies were inherently aligned with particular politics over others. The data in India is emerging from fieldwork at a crucial moment of scientific and medical history making. Here a very obvious and particular biocrossing occurred that allows for a nuanced understanding of contemporary and future articulations of cellular therapies and research. Patients participated in and enabled this biocrossing from embryonic to autologous just as much as policymakers and physicians. Discourse emerging from and managed by media, policymakers, and particular interest groups over others had a crucial role in promoting this biocrossing rather than a purely scientific evaluation. A cyclical relationship evolved where patients wanted to gain access to cutting-edge biomedical interventions; however, they were made cognizant of the possible risks/dangers and thus refrained from being 'too experimental'. What may be naturalized as patient preference for autologous somatic cells is far from a natural or symbiotic process but is rather a carefully constructed and politically motivated paradigm of permissible science.

Conclusion: Biocrossing Utopias and Dystopias

This 'biocrossing' from embryonic to somatic sources of cells is only the latest development in the stem cell research and therapy heterotopia. The heterotopia is indicative of a conceptual space in which cellular cultures gestate in contemporary India. It shows in no uncertain terms that cellular science is far from stabilizing anytime soon and in a perpetual state of movement and crossing(s) onto other terrains; however, to look at the factors and impetuses of these movements and biocrossings allows us to lay bare the political, economic, and ethical forces that attempt to naturalize (and perhaps de-politicize) one form of cellular medicine over others. However, there are notable exceptions to these moves, as evidenced by the presence of hESC therapy in India and emerging biographies of global patients embodying these cells (Bharadwaj 2013).

Perhaps, by either utilizing or building on biocrossing as a conceptual term, we can account for the emerging reality of stem cells as a global heterotopia when viewed from a vantage point that is uniquely Indian. The efforts to establish and partake in a globalized research system are leading the Indian state to prefigure the field in very particular ways. Rather ironically, hESC therapies in India are establishing a global presence attracting therapeutic citizens from around the globe to partake in a cellular breakthrough being ostracized in some quarters (see Bharadwaj 2015). This irony only enlarges the scope of biocrossings on a global scale. As Foucault observed as part of his third principle on heterotopias, a 'heterotopia is capable of juxtaposing in a single real place several spaces, several sites that are in themselves incompatible' (1967, p. 25). The contingent and context-sensitive ordering within a heterotopia is a fertile ground for assembling incompatible compatibles that continually reflect and refract the politics of making and unmaking.

References

Bharadwaj, Aditya. 2008. Biosociality and Biocrossings: Encounters with Assisted Conception and Embryonic Stem Cell in India. In *Biosocialities, Genetics and the Social Sciences: Making Biologies and Identities*, ed. Sahra Gibbon and Carlos Novas, 98–116. London; New York: Routledge.

———. 2012. Enculturating Cells: The Anthropology, Substance, and Science of Stem Cells. *Annual Review of Anthropology* 41 (1): 303–317.

———. 2013. Subaltern Biology? Local Biologies, Indian Odysseys and the Pursuit of Human Embryonic Stem Cell Therapies. *Medical Anthropology* 32 (4): 359–373.

————. 2015. Badnam Science? The Spectre of the 'Bad' Name and the Politics of Stem Cell Science in India. *South Asia Multidisciplinary Academic Journal.* https://samaj.revues.org/3999. Accessed 31 August 2016.

Central Drugs Standard Control Organization. 2013. Guidance Document for Regulatory Approvals of Stem Cell and Cell Based Products (SCCPs)/SPS/2013-001. New Delhi: Ministry of Health and Family Welfare. http://www.cdsco.nic.in/writereaddata/DRAFT%20GUIDANCE%20STEM%20CELLS-FINAL.pdf. Accessed 31 August 2016.

Foucault, Michel. 1967. "Of Other Spaces," Trans. Jay Miskowiec. *Diacritics* 16 (1): 22–27.

Franklin, Sarah. 2006. Embryonic Economies: The Double Reproductive Value of Stem Cells. *BioSocieties* 1 (01): 71–90.

————. 2007. Stem Cells R Us: Emergent Life Forms and the Global Biological. In *Global Assemblages*, ed. Aihwa Ong and Stephen J. Collier, 59–78. Oxford: Blackwell Publishing Ltd. http://onlinelibrary.wiley.com/doi/10.1002/9780470696569.ch4/summary. Accessed 31 July 2014.

————. 2013. *Biological Relatives: IVF, Stem Cells, and the Future of Kinship.* Durham, NC: Duke University Press. http://www.oapen.org/search?identifier=469257. Accessed 11 February 2016.

Hetherington, Kevin. 1997. *The Badlands of Modernity: Heterotopia and Social Ordering.* London: Psychology Press.

Indian Council for Medical Research. 2007. *Guidelines for Stem Cell Research and Therapy. Guideline Document.* New Delhi: Department of Biotechnology and Indian Council of Medical Research. http://icmr.nic.in/stem_cell/stem_cell_guidelines_2007.pdf. Accessed 20 April 2017.

————. 2013. *National Guidelines for Stem Cell Research. Guideline Document.* New Delhi: Department of Health Research and Department of Biotechnology. https://www.ncbs.res.in/sites/default/files/policies/NGSCR%202013.pdf. Accessed 18 April 2017.

Jayaraman, K.S. 2014. Unproven Stem Cell Therapy Banned. *Nature India.* http://www.natureasia.com/en/nindia/article/10.1038/nindia.2014.39. Accessed 18 April 2017.

Rajagopal, Divya, and Rohini Mohan. 2015. India's Disproportionately Tiny Health Budget: A National Security Concern? *The Economic Times*, October 31. http://economictimes.indiatimes.com/industry/healthcare/biotech/healthcare/indias-disproportionately-tiny-health-budget-a-national-security-concern/articleshow/49603121.cms. Accessed 18 April 2017.

Sleeboom-Faulkner, Margaret, and Prasanna Kumar Patra. 2009. The Bioethical Vacuum: National Policies on Human Embryonic Stem Cell Research in India and China. *Journal of International Biotechnology Law* 5 (6): 221–234.

Street, Alice, and Simon Coleman. 2012. Introduction: Real and Imagined Spaces. *Space and Culture* 15 (1): 4–17.

Thompson, Charis. 2013. *Good Science: The Ethical Choreography of Stem Cell Research*. Cambridge, MA: The MIT Press.

Tiwari, Shashank S., and Sujatha Raman. 2014. Governing Stem Cell Therapy in India: Regulatory Vacuum or Jurisdictional Ambiguity? *New Genetics and Society* 33 (4): 413–433.

Winner, Langdon. 1980. Do Artifacts Have Politics? *Daedalus* 109 (1): 121–136.

Nayantara Sheoran Appleton is Lecturer in Cultural Anthropology at Victoria University of Wellington, New Zealand. Prior to this post, she was a postdoctoral fellow in the Department of Anthropology and Sociology at the Graduate Institute of International and Development Studies, Geneva. Her research and teaching areas of interest include feminist medical anthropology and science and technology studies (STS); cultural studies and media anthropology; reproductive and contraceptive justice; ethics and governance; regenerative medicine; and ethnographic research methodologies. Her interest in biomedical interventions and burgeoning biotechnologies in contemporary India is reflected in her two key projects. First of these is on the feminist politics of health and reproduction in liberalizing India and second on the regulatory and ethical implications of emerging stem cell biotechnologies. Her dissertation research and writing has been supported by the National Science Foundation's (NSF) Doctoral Dissertation Research Improvement Grant in the Science, Technology, and Society (STS) Program (Award No: 1026682) and also the Dean's Dissertation Completion Fellowship at George Mason University. Her postdoctoral research fellowship was supported under a grant from the European Research Council (ERC) (Project ID: 313769, PI: Dr. Aditya Bharadwaj) titled "Red Revolution: The Emergence of Stem Cell Biotechnologies in India."

Aditya Bharadwaj is a professor at the Graduate Institute of International and Development Studies, Geneva. He moved to the Graduate Institute, Geneva, in January 2013, after completing seven years as a lecturer and later as a senior lecturer at the School of Social and Political Studies at the University of

Edinburgh. He received his doctoral degree from the University of Bristol and spent more than three years as a postdoctoral research fellow at Cardiff University before moving to the University of Edinburgh in 2005. His principal research interest is in the area of assisted reproductive, genetic, and stem cell biotechnologies and their rapid spread in diverse global locales. In 2013, he was awarded a European Research Council Consolidator Grant to examine the burgeoning rise of stem cell biotechnologies in India. Bharadwaj's work has been published in peer-reviewed journals such as *Medical Anthropology; Ethnos; BioSocieties; Social Science & Medicine; Anthropology & Medicine; Health, Risk & Society;* and *Culture, Medicine and Psychiatry* and he has contributed several chapters to edited collections. He has co-authored *Risky Relations: Family, Kinship and the New Genetics* (2006) and is the lead author of *Local Cells, Global Science: The Proliferation of Stem Cell Technologies in India* (2009). His sole-authored research monograph is titled *Conceptions: Infertility and Procreative Technologies in India* (2016).

Afterword

Marcia C. Inhorn

"Bio-crossings"—a term coined by editor Aditya Bharadwaj—is an apt neologism for this remarkable volume, which focuses on a miniscule biological entity, the stem cell, and the momentous ways in which this once-inconsequential bio-form has now become a powerful bio-technology, touching the lives of many people across the globe. These include scientists, clinicians, regulators, policymakers, patients, parents, advocates, anthropologists, and sociologists—many of whom have crossed national borders and disciplinary boundaries in their pursuit of stem cells as a new field of science and discovery, a powerful cure for some of the world's most dreaded afflictions, and a platform for regenerative medicine in the twenty-first century.

Yet, rarely do these stem cell stakeholders come together, which is exactly why this volume is so path-breaking and important. This book emerged from a one-of-a-kind, international conference held in Geneva, Switzerland, and called "Intersections: Social Science & Bioscience Perspectives on Stem Cell Technologies." The goal of that unique conference was to bring together five major groups of stakeholders:

© The Author(s) 2018
A. Bharadwaj (ed.), *Global Perspectives on Stem Cell Technologies*,
https://doi.org/10.1007/978-3-319-63787-7

1. Scientists who produce stem cell research, mostly but not exclusively in the global North;
2. Clinicians who use stem cells to treat sick patients, almost always in the global South;
3. Sick patients and their supporters, primarily their parents, who seek stem cell treatments for debilitating conditions;
4. Professional associations and their representatives, who attempt to provide ethical guidance and regulatory oversight; and
5. Social scientists who are charting the "social life" of stem cells as they reach many different corners of the globe.

This book thus reflects these various perspectives. It begins with fundamental insights made by two of the world's leading technoscience scholars, Charis Thompson of the University of California, Berkeley, and Sarah Franklin, of Cambridge University. In her chapter, Franklin reminds us that in vitro fertilization (IVF), which was invented and introduced in England in the late 1970s, has become the "platform" technology for much that has followed (Franklin 2013). Human embryonic stem cell (hESC) lines were first created from excess embryos in IVF labs. To use Franklin's excellent analogy, IVF was the original "steam engine" in the reprogramming of reproductive biology, leading to the development of many other forms of "biological equipment."

However, as Franklin also notes in her chapter, these powerful "bio tools" have also created much controversy and "technological ambivalence." In the US, for example, early stem cell debates centered around America's abortion politics and the religious disagreements over the status of the human embryo. Thus, ethical restrictions on stem cell research quickly emerged during the George W. Bush years, reflecting a "Christian right" view of morality and ethics. However, as Thompson argues based on her many years as a social scientist observing the US stem cell sector (Thompson 2013), "good science" can only emerge from a more capacious view of ethics and a "multi-vocal," democratic, and participatory scientific process. In other words, Thompson asks us to reconsider what constitutes good science and good ethics.

This very question—what constitutes good, ethical stem cell science—is at the heart of this volume. One chapter by Linda Hogle takes up this question by examining standards of evidence. In 2015, a policy document emerging from President Barack Obama's White House argued for a more "fluid," flexible approach to biomedical innovation, to undo some of the more cumbersome aspects of the US regulatory process. However, as Hogle shows, such flexibility is difficult to achieve under US standards of evidence-based medicine (EBM), which has become established in the West as an "organizing principle" for assessing new medical technologies. In order to meet the standards of EBM, clinical efficacy and safety must be proven through large-scale, randomized clinical trials (RCTs). In the West, RCTs are now sine qua non for "proving" both safety and efficacy. However, these large-scale trials, employing computerized "big data," are difficult and expensive to produce, and are often funded by pharmaceutical companies, creating the potential for conflict of interest. Thus, the question remains: Are RCTs always necessary? Or, are other forms of evidence, such as those based in clinical practice, also useful?

RCTs also pose ethical quandaries that are questioned in this volume. For example, several of the authors question the validity of placebo-controlled trials, when giving a placebo to a very sick patient seems unethical, even morally unconscionable. Much is at stake for very sick patients, when they have no other options. Thus, Thompson calls attention to "ethical choreographies"—or the ways in which different stakeholders come to understand and enact ethics. The assumption that only one moral universe exists based on four ethical principles—respect for autonomy, nonmaleficence, beneficence, and justice—may be inadequate in capturing what is at stake and what really matters in people's everyday local moral worlds (Kleinman 2006).

This is an important insight when it comes to stem cell medicine, especially as it is being practiced in Asia. In India, but also in China (Song 2017), therapeutic stem cells have been deemed a "breakthrough" technology—a veritable "revolution" in the treatment of otherwise chronic, incurable diseases and injuries. Such Asian centers of stem cell therapy are attracting stem cell "tourists" from around the world, including many from America and Europe. It is important to emphasize that in the West, stem cell therapy is not available for clinical use, for as described above, the evidentiary standards required to prove safety and efficacy through RCTs have not yet been established. In the US and Europe, then, stem cells constitute a kind of distant mirage—a treatment "on the horizon," but still many years down the road. Given this foreclosure, many sick Western patients—who have nothing else to lose and potentially much to gain—now make heroic journeys, often wheelchair bound, to stem cell clinics in Asia.

These Asian stem cell scientists and practitioners have been largely discredited by the international scientific community. For example, the chapter by Marcia Middlebrooks and Hazuki Shimono explores the "scandals" that took place in South Korea and Japan involving two separate stem cell scientists. Such individuals have been cast by the international community as "rogue" scientists or charlatans. Furthermore, attempts within different Asian countries to actually treat patients with stem cells are cast as "quackery," or the commercial exploitation of those who are desperate and suffering.

However, most of the chapters in this volume chart a far different Asian story, one that belies this kind of scientific disbelief and the frank paternalism on the part of the Western biomedical community. Part II, "Therapeutic Horizons" and Part III, "Patient Positions" focus specifically on India, a nation that has gained an international reputation—if not a scientifically accredited one—as the world's global stem cell "hub." As shown by Appleton and Bharadwaj in their major study of India's stem cell industry, clinics offering therapeutic stem cell treatments now operate all over India, in the major tier-one cities such as Delhi and Mumbai, as well as smaller, provincial, tier-two cities. There, physicians offer stem cell treatments for conditions ranging from rheumatism to optic nerve damage to amyotrophic lateral sclerosis (ALS), the deadly disease also known as Lou Gehrig's disease and made famous by the case of Nobel-prize-winning physicist Stephen Hawking.

Part II begins with the detailed work of Dr. Geeta Shroff, India's most well-known stem cell physician, who has been offering patients treatment since the early 2000s. Shroff charts the clinical history of her therapeutic stem cell line, which was derived from a single human embryo. Over the past decade and a half, Shroff has treated hundreds, if not thousands, of patients, many of them coming from Western countries, often with incurable, degenerative conditions. In her chapter, Shroff describes the treatment she offers to patients suffering from spinal cord injuries (SCIs) and the sometimes subtle, but often substantial improvements that she has documented in both quadriplegic and paraplegic patients. In the next chapter, Dr. Petra Hopf-Seidel, a German psychiatrist and neurologist, describes the journeys she has made to Dr. Shroff's clinic accompanying her very sick German patients, including those suffering from the neurological effects of chronic Lyme disease and ALS. Among the Lyme patients but less so among the ALS sufferers, Hopf-Seidel has seen some remarkable improvements, even full recovery. Thus, she continues to advocate stem cell therapy for her neurologically impaired patients, especially those who have few other treatment options.

The final section of this book, Patient Positions, is, in my view as an anthropologist, the most powerful and thought-provoking. Taking patient subjectivities and voices very seriously, Bharadwaj sought in this volume to include the perspectives of those who have actually sought out and used stem cell therapies in India. Thus, in this section, one chapter features an interview with a mother and daughter. Shannon Davis, a young American woman, was rendered quadriplegic by a horrifying auto accident. Her mother Lola did everything she could to support Shannon. Through research on the internet, Shannon and Lola discovered that their only hope for stem cell therapy was in India with Dr. Shroff. Thus, they made the trip together several times for stem cell treatments, which were delivered via injection and spinal infusion. Once immobile from her chest down, Shannon can now function on her own. She feeds herself, dresses herself, moves herself from her wheelchair into her specially adapted car, and drives herself to work, where she is employed in the job that she held before the accident. The Davis' mother-daughter conversation serves as a powerful testimonial to the regenerative efficacy of stem cell therapy, which in the case of Shannon has led to vast improvements in her quality of life.

In the moving chapter by Ripudaman Singh, he recounts how he became a parent activist, when his young son was diagnosed with the degenerative and ultimately deadly condition known as Duchenne muscular dystrophy (DMD). Heartbroken but determined to do something for his beloved son, Singh becomes a kind of "lay expert" in stem cell therapy. Along with several other Indian DMD parents, Singh and his compatriots self-fund the Adipose Stem Cell In-Vitro Lab Study, in which they enroll their sons (because DMD is a sex-linked genetic condition, primarily affecting boys). Although Singh faces criticism for turning his son into an experimental subject, he is supported by wife, his family, and his physician in-laws. To date, his son has suffered no ill effects, although also no marked improvements. As Singh explains so poignantly, without these kinds of "patient-driven" studies, there is literally no hope for DMD patients in India. Singh laments what he sees as the lack of concern for the disabled: "In India, the reality is that nobody cares about a disabled child, and even if I were to narrate my story, nobody cares. It is ironic that India is a spiritual country but it has so much apathy."

Thus, the heartbreak and desperation, the hopes and fears, the need for compassion and for evidence, and ultimately the desire for a cure ring out in this final poignant section. Unlike so many other ethnographic volumes, in which social scientists speak "for" their interlocutors, this volume is remarkable in letting patients and their parents speak for themselves. It reveals their struggles, their heartbreaks, their desires, and their hopes. These chapters attest to the fact that patients and their advocates *must* be part of the stem cell conversation.

Finally, it is important to end on the theme of social justice. In a country like India, where poverty is rampant and the public health system is broken down, stem cell therapy exists in the world of private, fee-for-service medicine, where patients must pay, sometimes high prices, for stem cell services. It is thus not surprising that many of the beneficiaries of stem cell treatments in India are arriving from the global North. Or, like the Indian parents who funded the DMD study, they are educated elites within their own society. The advent of stem cell therapies in India and other Asian settings raises thorny questions about healthcare access and social justice, including how patients arriving from the global North may benefit at the expense of those from the global South. In an era when stem cell therapies are still globally inaccessible, questions of prioritization, triage, equity, and justice become paramount.

Although this book is being published decades after the first stem cell lines were established in the West, therapeutic stem cells are far from being offered in Western clinical practice. Indeed, the slow pace of development in the West has led to other global trajectories, intersections, and perspectives. This book has admirably captured this lively and important domain of biomedicine outside of the West in a part of the world that is showing itself to be a lively site of technoscientific invention. The book is timely and thought-provoking foray into this world, and the global circulations that are making it possible.

Anyone interesting in stem cells should read this book.

For anyone interested in stem cell treatment, reading this book is a must.

References

Franklin, Sarah. 2013. *Biological Relatives: IVF, Stem Cells, and the Future of Kinship*. Durham, NC: Duke University Press.

Kleinman, Arthur. 2006. *What Really Matters: Living a Moral Life Amidst Uncertainty and Danger*. Oxford, UK: Oxford University Press.

Song, Priscilla. 2017. *Biomedical Odysseys: Fetal Cell Experiments from Cyberspace to China*. Princeton, NJ: Princeton University Press.

Thompson, Charis. 2013. *Good Science: The Ethical Choreography of Stem Cell Research*. Cambridge, MA: MIT Press.

Index[1]

---

[1] Page numbers followed by "n" refer to notes.

© The Author(s) 2018
A. Bharadwaj (ed.), *Global Perspectives on Stem Cell Technologies*,
https://doi.org/10.1007/978-3-319-63787-7